A FIRST COURSE IN
ORGANIC CHEMISTRY

X-ray structure on cover taken from G. Ferguson, B. Kaitner, D. Lloyd and H. McNab (1984). *J. Chem. Res.* (S) **182**, (M) 1738. See also p. 320 of the present text.

A FIRST COURSE IN ORGANIC CHEMISTRY

DOUGLAS LLOYD

Department of Chemistry, University of St. Andrews, St. Andrews, Fife, Scotland

JOHN WILEY & SONS

Chichester · New York · Brisbane · Toronto · Singapore

Other Wiley Editorial Offices

John Wiley & Sons, Inc., 605 Third Avenue,
New York, NY 10158–0012, USA

Jacaranda Wiley Ltd, G.P.O. Box 859, Brisbane,
Queensland 4001, Australia

John Wiley & Sons (Canada) Ltd, 22 Worcester Road,
Rexdale, Ontario M9W 1LI, Canada

John Wiley & Sons (SEA) Pte Ltd, 37 Jalan Pemimpin #05–04,
Block B, Union Industrial Building, Singapore 2057

British Library Cataloguing in Publication Data:
Lloyd, Douglas
  A first course in organic chemistry.
  1. Organic compounds
  I. Title
  547

  ISBN 0 471 92408 3
  ISBN 0 471 92409 1 pbk

Typeset by Thomson Press (I) Ltd., New Delhi.
Printed in Great Britain by Biddles Ltd., Guildford.

CONTENTS

PREFACE

Oh, not another new textbook on organic chemistry! I am afraid it is. For goodness' sake, why? And what is there new to say?

In teaching an introductory course on organic chemistry to first-year students in recent years it has seemed to me that none of the texts, some of them excellent, quite presents the feel of organic chemistry *as it is presently done.* Since the best teaching is by a practitioner instructing the apprentice, it seems evident that one should communicate a subject as it is practised. Above all in the case of organic chemistry, this means treating the use of spectroscopic techniques, and especially n.m.r. spectroscopy, as routine practice, and as a normal part of the chemistry of compounds, rather than as a special topic. The use of n.m.r. is therefore introduced at the start, in the consideration of the structure of alkanes, and then used whenever apposite throughout the book. Concise accounts of different physical techniques, which are now an everyday part of organic chemistry, are given in an appendix, so that they do not interfere with the flow of the chemistry in the main text.

The text is aimed principally at first-year students at universities, polytechnics or colleges, or senior pupils at schools, who have already been provided with some general knowledge of chemistry. I assume, for example, that the reader already knows about atoms and electrons. However, the organic chemistry starts from its beginning. I have tried to develop the subject logically, giving an understanding of molecular structure and its elucidation, and of reactions and how they proceed.

Most organic chemists are now agreed, I think, that the functional group approach is the better method of introducing organic chemistry and classifying it clearly, and this is the method I follow.

The order of presentation has some unusual features, which have their reasons, usually because of links established between adjacent subjects. The structure of alkanes is followed by the structure and reactions of alkenes, and then come the reactions of alkanes. One must start consideration of organic chemistry and the structure of organic molecules with alkanes; they are the parent compounds. However, their chemistry is rather atypical of the commoner elementary organic reactions, being bereft of ionic reactions. There is some advantage in introducing organic reactions via the commoner ionic type of reaction, and then going on to the radical reactions of alkanes. As well as being the type of reaction that pervades most of this book, ionic reactions are also probably easier to explain and grasp than are radical reactions, and this also prompts their being mentioned first.

Some compounds not usually dealt with in elementary textbooks of organic chemistry are described. An example is provided by cyclic ethers. It is remarkable that so simple a compound of such industrial importance as ethylene oxide, of which millions of tons are made every year, is often totally ignored or given scant attention.

The concepts of orbitals, hybridization or σ- and π-bonding are not introduced. At the level of this text these relatively complex concepts are not necessary and may serve to confuse rather than to clarify. For students studying chemistry further these aspects can be added afterwards, without any problems or difficulties.

Throughout the text I make continuing reference to the importance of organic chemistry in everyday life, in particular to the role and use of organic compounds in industry, in everyday life and in biology. The study of organic chemistry could be justified solely on its importance in everyday life, but a proper appreciation of this importance can only come with a prior knowledge of the fundamentals of organic chemistry. Organic chemistry is organic chemistry whether it takes place on the laboratory bench, in a chemical factory, or in plants or bodies, including human bodies. It therefore seems more real to make reference to these industrial and biological aspects alongside the fundamental academic chemical aspects, at the relevant places, rather than to hive them off into separated chapters as though they were somehow different.

Even greater than this relevance to everyday life as a reason for studying organic chemistry is the satisfaction, both intellectual and aesthetic, that can be derived from its study, not to mention a concomitant training in logical thinking. How nice to have a subject that is both fun and intellectually satisfying, and at the same time is so important to all of us and is used, wittingly or unwittingly, by **all** of us **all** the time.

If this book can provide the reader with some sense of this pleasure and intellectual stimulus, give some idea of the flavour of contemporary organic chemistry, and also open the reader's eyes to the literally vital role of organic chemistry in the world around us, so that he/she has a richer appreciation of the surroundings than if no organic chemistry was known, then the labour of writing it will be amply rewarded.

I must thank a number of people for their kind help in different ways. I thank my wife (who is not a chemist), and my friends Dr Hamish McNab (Edinburgh), Dr Ray Mackie and Dr Christopher Glidewell (St Andrews) for reading parts of the text and providing invaluable constructive criticism. Dr Shirley Seth and Mrs Melanja Smith kindly produced the n.m.r. spectra which appear in the text. Professor George Ferguson (Guelph, Ontario) recorded the X-ray structure illustrated in the Appendix. A number of other friends, especially Dr Gordon Harris and Dr John Walton, also made most helpful comments on various points, which I have been glad to incorporate. Finally I must thank the staff of John Wiley & Sons for their kind interest and help throughout the preparation of the book.

St Andrews, Fife.
i.89.

1 ORGANIC CHEMISTRY

Why is there a division of chemical compounds into 'organic' and 'inorganic'? The names evolved from the history of chemistry. Early chemists obtained their compounds from natural sources, some animate, some inanimate. Those compounds derived from material which was or had been alive, such as plants and animals, came to be called organic; those from inanimate matter such as rocks were called, by contrast, inorganic. Indeed, at one time there was thought to be a fundamental difference between organic and inorganic materials. However, it was realized that there was no real fundamental difference; all were made up from associations of atoms.

It was also realized that almost all the so-called organic compounds contained carbon and the use of the terms organic and inorganic took on the meaning of compounds containing carbon (= organic) and compounds not containing carbon (= inorganic). Even this definition has its exceptions. For example metal carbonates and bicarbonates contain carbon, but are usually included in inorganic chemistry. In general, however, the accepted definition of organic compounds is that they are compounds containing carbon.

It may now be asked why, when there are over a hundred elements, the study of just **one** of those elements, carbon, and the compounds it forms, is separated from that of the remaining elements. This separation is a matter of convenience, based on the fact that a very large proportion of all reported compounds are carbon compounds. Thus it is not inappropriate to consider these carbon compounds as a group separate from other compounds. It needs to be emphasized that there is no real distinction between organic and inorganic compounds; all have to follow the dictates of the same chemical laws. It is nonetheless convenient to consider this large group of carbon compounds separately; organic chemistry is still, however, but one chapter, even if a rather large one, in the overall story of chemistry.

Organic Chemistry In and Around Us

Organic chemistry is, in the most literal sense of the word, vital. The maintenance of life in plants and animals, including ourselves, is utterly dependent upon millions of organic chemical reactions taking place continuously within ourselves and in all other forms of living matter. A chemical reaction that goes wrong in a laboratory may be a nuisance to the worker or may even cause a disaster. A chemical reaction within our bodies that goes wrong may also prove disastrous, to the person concerned, for it may result in disease or death.

Organic reactions sustain us daily. Such reactions enable the digestion of our food to provide the building materials for our bodies and the energy to keep them going. Other organic reactions have provided the food we eat.

It is not only in the processes of life that organic reactions concern us. When we burn fuel to provide heating or transport we are using controlled organic reactions. Many of the products which we consider necessary for everyday life, such as plastics, drugs, dyestuffs, synthetic fibres for clothing and other uses, are organic compounds prepared by controlled organic reactions.

Even death and decay continue the process; again organic reactions are involved!

Why can Carbon Form so Many Compounds?

Carbon is unique among the elements in that carbon atoms can form stable bonds with one another to form chains and/or rings of many atoms linked together. These organic molecules can be made up from an inifinitely variable number of carbon atoms, which can in turn be linked together in an equally variable series of patterns, for the chains and rings may (or may not) have branches, of unlimited variety, attached to them.

What properties must an element have to be able to provide these complex molecular structures?

First it must have a **covalency*** greater than two. This is essential so that branching of a chain or ring of atoms is possible. An element X with a covalency of two could only have four of its atoms arranged in two ways in molecules:

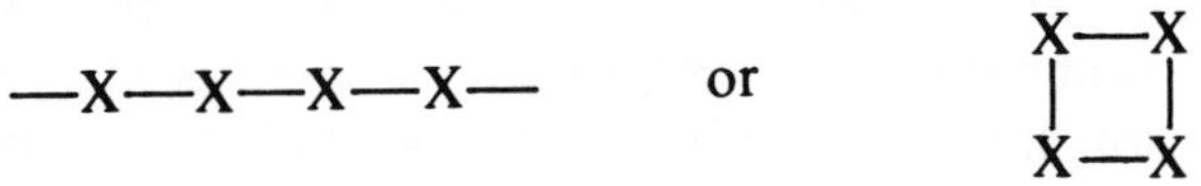

$$-X-X-X-X- \qquad \text{or} \qquad \begin{array}{c} X-X \\ |\quad\ | \\ X-X \end{array}$$

Carbon has a covalency of four. Hence four carbon atoms in a molecule could be arranged in five ways, as follows:

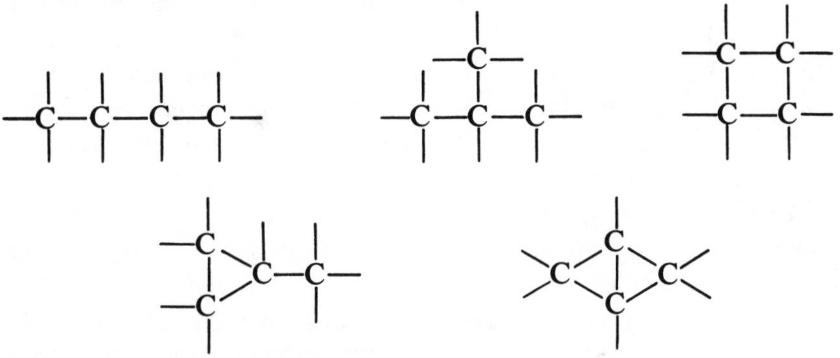

In these structures not all of the four **bonds** of the different carbon atoms are used in order to join the carbon atoms together, and no account is taken of what other atoms are attached to the skeleton of carbon atoms. We are only considering here the complexity that may arise in the arrangement of the carbon atoms themselves. As the number of carbon atoms in a molecule is increased, so is the complexity of their possible arrangement. It is thus not surprising that very many different carbon compounds are known. For example, if only structures not having rings are considered, there are more than 300,000 ways of arranging twenty carbon atoms!

The second important factor is that the bonds linking the atoms together are energetically stable i.e. strong enough at room temperature not to break. Additionally they must be reasonably resistant to chemical attack by ever-present materials such as water or oxygen. They must also be resistant to cleavage brought about by the action of light.

*'Covalency' is used here to mean the number of covalent bonds that an atom of the element usually forms.

Of all the elements carbon alone meets all these criteria completely. We can at once rule out all monovalent and divalent elements, and all metallic elements, which do not form stable covalent bonds with themselves in this way.

This really leaves only carbon, nitrogen, phosphorus, boron and silicon for consideration. The energy required to break a C—C bond is 350 kJ/mol, which is not readily available in ambient conditions. To break a N—N bond, on the other hand, requires only 160 kJ/mol; this value is too low to provide the necessary stability in all but the simplest structures. In consequence molecules having chain of nitrogen atoms are rare. B—B and Si—Si bonds are similarly weak, i.e. they are readily broken. The P—P bond is somewhat stronger in that it needs 215 kJ/mol to break it, but in this case such bonds are easily broken by reaction with water; they undergo what is called **hydrolysis.** Hence compounds having such bonds have little chance of surviving long in normal conditions.

Most organic molecules have, in addition to C—C bonds, C—H bonds. The energy required to break C—H bonds is ∼ 410 kJ/mol, i.e. in organic molecules containing carbon and hydrogen a strong C—C framework is surrounded by hydrogen atoms linked by strong bonds to the carbon framework. Compounds containing **only** hydrogen atoms attached to the carbon framework are called **hydrocarbons.**

It may be noted that a C—F bond is even stronger than a C—H bond; it requires ∼ 440 kJ/mol to bring about its cleavage. It is not surprising therefore that there is a family of molecules containing only fluorine atoms attached to the carbon skeleton, called **fluorocarbons**, which are exceedingly stable, and, in consequence, find much commercial application, e.g. for making chemically and heat resistant parts. In ordinary domestic life they are found in most kitchens, providing the non-stick layer applied to frying pans.

The Study of Organic Chemistry

Organic chemistry deals with the study of carbon compounds. For the reasons given earlier there are millions of such compounds. Despite this it is possible to study them in a systematic way. Two general features can be considered: the structure of the compounds and the reactions in which they participate; not surprisingly the latter—how the compound can behave chemically—is connected with the former—how the compound is made up from its constituent atoms. Furthermore, these studies involve not just the academic study of chemistry for the sake of increasing our knowledge of the subject, but also its all-pervasive role in everyday life, both in biological activity and in other aspects of living and in industry.

The everyday importance of organic chemistry will be emphasized throughout this text, which must of necessity, however, as a teaching and learning text, base its approach on academic lines. The main interests of organic chemistry can be listed as follows:

Determination of the structure of molecules
Correlation of structure and properties of compounds
Synthesis/purification of organic compounds
The study of reaction mechanisms

Each of these topics is now considered severally.

Determination of the structure of molecules

This requires the discovery of (a) the constituent atoms from which any organic molecule is made up, i.e. what elements other than carbon are also present in the molecule, (b) the numbers of each of the atoms of the constituent elements that are present, (c) the order in which these constituent atoms are linked together in the molecule and (d) the shape of the resultant molecule.

The types and numbers of atoms in a molecule are readily determined by **elemental analysis**. For example, we can find the amount of carbon and hydrogen in a compound by completely burning a very small sample of the compound, say ~ 1 mg. In this process the carbon and hydrogen contained in the compound are completely converted into carbon dioxide and water respectively. From the amounts of these compounds that are formed, we can calculate the amounts of carbon and hydrogen in the original compound.

As an example let us consider the combustion of 4.6 mg of ethanol, the 'alcohol' of alcoholic beverages such as beer and wine. From this sample 8.8 mg of carbon dioxide and 5.4 mg of water are obtained.

Carbon dioxide has the formula CO_2. The atomic weights of carbon and of oxygen are, respectively, 12.01 and 16.00; the molecular weight of carbon dioxide is 44.01. Therefore of the 8.8 mg of carbon dioxide $(8.8 \times 12.01/44.01) = 2.4$ mg are carbon. Similarly, since the atomic weight of hydrogen is 1.01 and the molecular weight of water 18.02, it follows that the weight of hydrogen in the analytical sample was $(5.4 \times 1.01 \times 2/18.02) = 0.6$ mg. Since the weights of hydrogen and carbon contained in the ethanol are 2.4 and 0.6 mg it follows that the ratio of the numbers of hydrogen and carbon atoms, obtained by dividing the weights of each element by the respective atomic weights, is $2.4/12.01:0.6/1.01, = 1:3$.

Our analysis also shows that an element other than carbon and hydrogen must be present since 2.4 mg of carbon and 0.6 mg of hydrogen were contained in a sample weighing 4.6 mg. In fact this other element is oxygen, which is less commonly analysed for. If we divide the weight of oxygen present $[4.6 - (2.4 + 0.6)] = 1.6$ mg by the atomic weight of oxygen (16.00) we can again compare the ratio of the numbers of atoms present, viz.:

	Weight present (W)(mg)	W/atomic weight	ratio of numbers of atoms
C	2.4	$2.4/12.01 = 0.2$	2
H	0.6	$0.6/1.01 = 0.6$	6
O	1.6	$1.6/16.00 = 0.1$	1

This still does not show whether the **molecular formula** of ethanol is C_2H_6O or $C_4H_{12}O_2$ or $C_6H_{18}O_3$, etc. To distinguish between these possibilities it is necessary to measure the **molecular weight** (i.e. the sum of the atomic weights) of ethanol. This may be obtained by the use of mass spectrometry (see the Appendix). The molecular weight proves to be 46.08; hence the molecular formula must be C_2H_6O.

The next step is to find how these atoms are arranged in the molecule. Using the normal covalencies of carbon (4), hydrogen (1) and oxygen (2) two ways of arranging the constituent atoms are possible (X and Y):

$$\begin{array}{cc}
\text{H}\quad\text{H} & \text{H}\qquad\text{H} \\
|\quad\ | & |\qquad\ | \\
\text{H--C--C--O--H} & \text{H--C--O--C--H} \\
|\quad\ | & |\qquad\ | \\
\text{H}\quad\text{H}\quad(=C_2H_5OH) & \text{H}\qquad\text{H}\quad(=CH_3OCH_3) \\
\end{array}$$

(X)   (Y)

To distinguish between such possible structures, historically it was necessary to investigate their chemistry. Although this method may still have its uses, nowadays spectroscopic techniques are more commonly employed to find the **molecular structure**, i.e. the arrangements of the atoms within a molecule. As will be seen in later chapters, all such investigations on ethanol show unequivocally that it must have structure **X**.

Also in present practice it may prove possible to obtain both a molecular and a structural formula directly from mass spectra or by X-ray crystallography (see the Appendix).

To recapitulate, in determining the structure of a molecule it is necessary to discover

(a) what elements are present,
(b) the numbers of atoms of each element present,
(c) the way in which these atoms are arranged within the molecules.

In earlier times, when only chemical information could be available, structure determination could be a very lengthy process, taking years in the case of complicated molecules. The advent of modern spectroscopic techniques (see the Appendix) has enormously simplified and shortened structure determination, but it can still present difficulties.

Correlation of structure and properties of compounds

One important consequence of defining the chemical structure of organic compounds is to see if the various properties, both chemical and physical (and possibly physiological, etc.), of compounds can be associated with different structural features. Thus for a long time it has been recognized that the presence of hydroxy (HO—) groupings of atoms in molecules tends to increase the solubility of such molecules in water, and the presence of a carboxyl (—COOH) grouping makes molecules acidic. Such correlation is especially important with regard to spectroscopy, wherein information has been accumulated that absorption peaks at certain frequencies (wavelengths) are indicative of the presence of certain groupings of atoms. In recent times a great deal of attention has been applied to correlations between molecular structure and biological properties; elucidation of such correlations could be (and, indeed, already are) of enormous value in the design of drugs and of the understanding of biological processes.

Synthesis

The synthesis of an organic compound from simpler commonplace materials is first and foremost a stimulating challenge, both intellectual and practical. Probably its first main function was to help to prove the structures assigned to molecules. If a series of preparative steps, using reactions whose outcome could reasonably be assured, provided the compound required, then this provided strong evidence that the attributed structure was indeed correct.

On a more general level, if the structure of a naturally occurring compound, useful as, for example, a drug or a dyestuff, was discovered, then it might be worth while to try to devise ways of synthesizing this compound from readily available materials, so that bigger supplies, possibly also at lower prices, might become available. Furthermore, if it were thought that compounds related to useful naturally occurring compounds, which did not themselves occur in nature, might prove useful, then it was worthwhile to synthesize such compounds and examine their properties.

It is on such bases that the huge worldwide chemical industry has grown up, manufacturing either compounds that occur naturally, but in insufficient quantities, or totally new materials which can be useful for a wide variety of purposes. This industry provides us with a vast range of materials, plastics, dyestuffs, medicinals and many other different products, which most people have come to regard as essential ingredients of everyday life.

Purification of organic compounds

Perhaps the major skill required of an organic chemist is the ability to obtain pure samples of compounds. In academic work the structural analysis of a compound crucially depends on its purity. Presence of an impurity may lead to incorrect analyses and may give rise to spurious extra signals in spectra. In industrial processes impurities may have serious effects on the properties of the products. For example, the presence of even minute amounts of a toxic impurity in a pharmaceutical product could not be tolerated.

The classical methods of purification involved distillation and recrystallization—still methods of immense importance. Other techniques have been added, especially those utilizing chromatography in various ways for the separation of different compounds.

Detailed discussion of these techniques is outside the scope of the present text; details are readily obtained from the numerous textbooks of practical organic chemistry which are available. Even more valuable is to learn the practical feel of these techniques by performing them.

The study of reaction mechanisms

It is often easy to demonstrate that different reactions take place, e.g. to show that petrol burns. But how exactly do the molecules taking part in a reaction interact with each other? By well-designed experiments using appropriately chosen molecules, it is often possible to obtain quite detailed information about the reaction processes. Some examples will be mentioned in later chapters. In addition to providing more information and understanding of how chemical reactions take place, this understanding may also suggest ways by which a reaction may be made to go more efficiently. Therefore this study of **reaction mechanisms** may not only be of intellectual interest but may also assist synthetic processes and thus be of industrial value as well as academic.

This book will deal with examples of all the different aspects of organic chemistry mentioned above. The structure, and the determination of the structure of organic compounds, their synthesis, their reactions, and the mechanisms of many of these reactions, will all be considered and exemplified. In particular an attempt will be made throughout to put the chemistry discussed in everyday context, and the wider interest of this chemistry in nature and in industry and in everyday life will be mentioned.

Finally the value of organic chemistry as a logical system will be presented.

There are two main virtues of the study of organic chemistry. First, it inculcates a method of logical approach to problems. The chemical logic required to deduce the structure of a molecule or to plan a synthesis of a compound offers excellent training in logical approach and can be of intense intellectual satisfaction. Second, as discussed earlier in this chapter, the knowledge of organic chemistry not only extends our horizons but is indeed essential if we are to try to appreciate the world in which we live. Not only are so many of the artefacts we use the results and products of organic chemistry, but all nature and we ourselves are living examples of organic chemistry in action. No twentieth (or twenty-first) century man or woman can be considered to be truly literate without at least some understanding of the workings of this all-pervasive subject. And the appreciation of all that happens around us is much deeper and much more satisfying with this understanding.

2 ALKANES AND THEIR STRUCTURE

The simplest organic molecule is **methane**, a gas that occurs naturally in the earth. It burns readily in air, and in consequence has been both beneficial and disastrous for mankind. In the latter role it is the major constituent of dreaded firedamp in coal-mines, and has been the cause of numerous fires and explosions; many men have lost their lives because of its chemical properties. In its beneficient role it is the main constituent of the natural gas which is used widely for fuel, both domestically and industrially.

Structure of Methane

Elemental analysis and its molecular weight show that methane has the molecular formula CH_4. It has been shown by chemical reactions and by spectroscopic studies (see also later in this chapter) that all the four hydrogen atoms are identical to one another, i.e., they are symmetrically arranged around the one carbon atom. This might be achieved in one of two ways: (a) by having a square planar molecule (**A**) with all the atoms in one plane, all the hydrogen atoms at equal distances from the central carbon atom and also at equal distances from one another or (b) by having the hydrogen atoms arranged around the carbon atom to form the apices of a tetrahedron (**B**), again with all the hydrogen atoms at equal distances from the carbon atom and from one another.

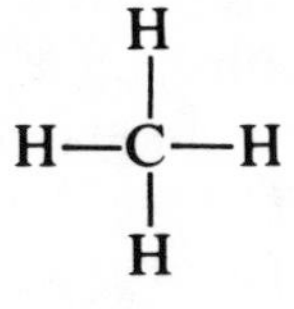

(A) 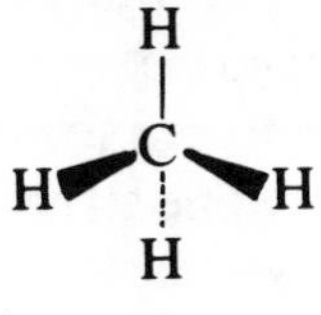   (B)

(Note that the lines joining the hydrogen atoms to the carbon atom are only indicating the spatial relationships of the atoms. Lines drawn as — in compounds **A** and **B** indicate that the atoms linked lie in the plane of the paper, a dotted line in compound **B** indicates that the hydrogen atom is behind the plane of the paper, while the wedge signs ◀ in **B** indicate that the hydrogen atoms next to them are above the plane of the paper; with such signs the thicker end of the wedge is always towards the reader.)

The decision about which of these structures is correct was initially the result of chemical studies.

Supposing a compound CH_3X is made in which one of the hydrogen atoms is replaced by another element, as in CH_3Br, bromomethane. Only one compound of this formula exists. Therefore all the four links to the central carbon atoms must be identical; if this were not so, different compounds could exist with the bromine atom attached to the different links. Structures such as either **A** or **B** would fit these results.

Now let us consider a compound such as CH_2X_2, for example CH_2Br_2, dibromomethane. Again only **one** such compound is known. This is consistent with a molecule having a tetrahedral shape, viz. structure **C**:

$$
\begin{array}{c}
H \\
| \\
H \blacktriangleright C \blacktriangleleft Br \\
| \\
Br
\end{array}
$$

(C)

The reader should convince him- or herself of this by making up a **molecular model** of compound C. Ideally this may be done using specially made molecular model kits. Alternatively a black blob of plasticine may be used to represent the carbon atom with four matchsticks of equal length stuck into the black blob to provide the appropriate tetrahedral shape. To the ends of two of these matchsticks, white blobs of plasticine may be added to represent the hydrogen atoms and to two other matchsticks two brown blobs of plasticine may be attached to represent the two bromine atoms. Whichever matchsticks are used for the two different coloured 'atoms' it will be found that only one structure can be made. (The specific use of black, white and brown plasticine is not necessary of course; any contrasting colours may be used, but it is the standard convention to represent carbon, hydrogen and bromine atoms by, respectively, black, white and brown pieces.)

By contrast, if the molecules had planar shapes as in structure **A**, **two** different compounds (**D** and **E**) would be possible, according to whether the two identical atoms are next to or opposite to one another. (The reader may well convince himself of this also by the use of molecular models.)

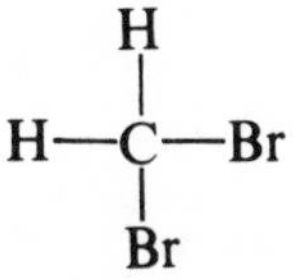

$$
\begin{array}{c}
\text{H} \\
| \\
\text{H—C—Br} \\
| \\
\text{Br}
\end{array}
\qquad\qquad
\begin{array}{c}
\text{Br} \\
| \\
\text{H—C—H} \\
| \\
\text{Br}
\end{array}
$$

(D) 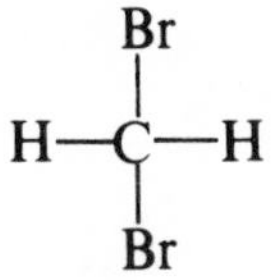   **(E)**

The arguments used in these paragraphs are typical of the logic and methodology used in organic chemistry. They were first presented as long ago as 1874, by the Netherlands chemist van't Hoff. This concept of the tetrahedral arrangement of four atoms about a carbon atom underpins the whole of organic chemistry. X-ray crystallography (see the Appendix), which enables one to obtain pictures of the arrangements of atoms within molecules, confirms this concept, which was originally deduced on the basis of chemical information, as described in the foregoing paragraphs.

Ethane

Ethane, C_2H_6, has two carbon atoms linked to one another, and to each of these carbon atoms are attached three hydrogen atoms. It may alternatively be written as CH_3CH_3, or $CH_3—CH_3$; a CH_3 group of atoms is known as a **methyl** group. The overall arrangement around each carbon atom is tetrahedral.

The molecule of ethane could be represented as follows (**F**):

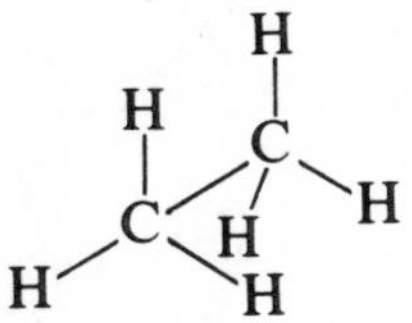

(F)

This is described as a 'sawhorse' type of formula and attempts to depict the shape of a molecule using only normal lines to connect atoms. Again a molecular model is helpful in appreciating what the molecule is like three dimensionally. It is usually possible to 'rotate' groups around a carbon–carbon single bond. Thus for ethane we could perform such a rotation of compound **F** through 180° and the molecule would then appear as follows (**G**):

$$
\begin{array}{c}
\text{H} \quad \text{H} \\
\diagdown \ / \\
\text{H} \quad \text{C} \\
| \quad / \\
\text{C} \quad \text{H} \\
/ \quad | \\
\text{H} \quad \text{H}
\end{array}
$$

(G)

Rotation of compound **F** through 60 or 120° would lead to the same shape as compound **G**.

This can also be demonstrated by viewing ethane in form **F** along the line of the carbon–carbon bond. For convenience the carbon atoms are shown as a black dot, the hydrogen atoms are omitted, the C—H bonds on the front carbon atom are shown as firm lines and the C—H bonds on the rear carbon atom as dotted lines. We then have:

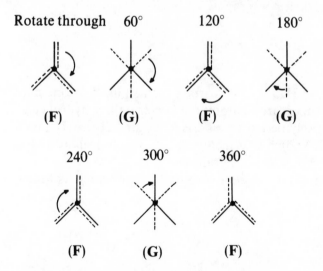

Rotate through	60°	120°	180°
(F)	(G)	(F)	(G)

240°	300°	360°
(F)	(G)	(F)

The different shapes, such as **F** and **G**, which a molecule can take up by rotation about a bond, are known as **conformations.**

When conformation **F** is viewed along the carbon–carbon bond, the carbon–hydrogen bonds on the front atom are in line with those on the rear atom. This is described as an **eclipsed** conformation. When conformation **G** is viewed in this way, the carbon–hydrogen bonds on the front and rear atoms are seen to be as far apart from one another as is possible. This is known as a **staggered** conformation. In addition to these extreme cases, the molecule can also take up any of the intermediate shapes between eclipsed and staggered.

Experiment shows that the staggered form is in fact more stable than the eclipsed form, by $\sim 12\,\text{kJ/mol}$. The change in stability of the molecule as the two constituent methyl groups are rotated with respect to one another may be depicted graphically as follows:

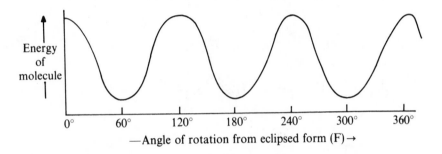

—Angle of rotation from eclipsed form (F) →

Note that the lower the energy of the molecule, the more stable it is. Hence its preferred shape is in a staggered conformation. Ethane probably spends a preponderant amount of time in its staggered conformation.

Rotation about the carbon–carbon bond occurs but is not completely free because the molecule requires energy to pass through the eclipsed conformations. However, at room temperature there is always ~ 60–$80\,\text{kJ}$ of energy available from the environment so complete rotation about the carbon–carbon bond is readily achieved. To inhibit such rotation completely the energy barrier involved would need to be ~ 80–$125\,\text{kJ/mol}$.

Alkanes

The formulae of both methane and ethane may be expressed in the general form C_nH_{2n+2}, where $n = 1$ in the case of methane and $n = 2$ in the case of ethane. Compounds made up of carbon and hydrogen atoms only are called **hydrocarbons.** There is a large group of compounds that all have the molecular formulae C_nH_{2n+2}, known generally as **alkanes.** The next members of this series, with $n = 3$ and 4, are, respectively, **propane,** C_3H_8, and **butane,** C_4H_{10}. Their structural formulae may be expressed as

$$CH_3-CH_2-CH_3 \quad \text{or} \quad CH_3CH_2CH_3$$
Propane

$$CH_3-CH_2-CH_2-CH_3 \quad \text{or} \quad CH_3CH_2CH_2CH_3$$
Butane

For propane there are eclipsed and staggered conformations, as in ethane, but this time a methyl group is eclipsed by a hydrogen atom as well:

Eclipsed

Staggered

In the case of butane there are **two** different eclipsed forms, in one of which two methyl groups eclipse one another, while in the other the methyl groups are eclipsed by hydrogen atoms, viz.

and 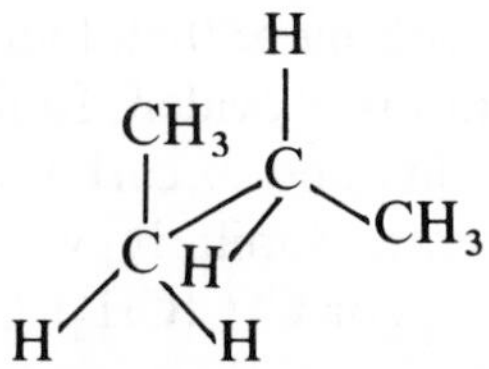 or 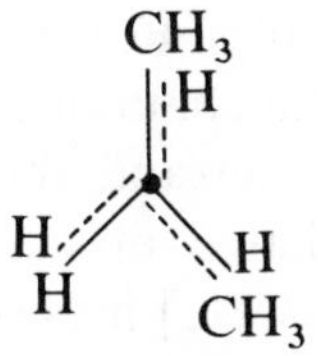

Similarly there are **two** staggered forms, one with two methyl groups at 60° to each other and the other with them at 180° to each other:

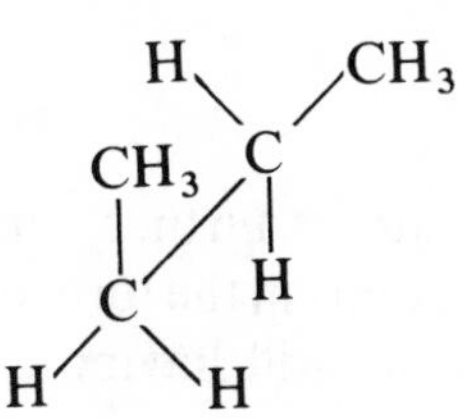 or 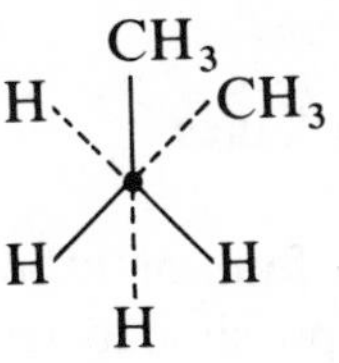

and 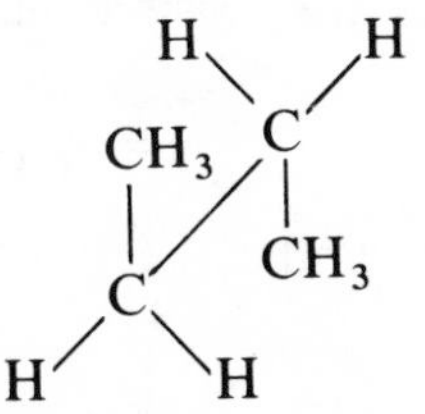 or 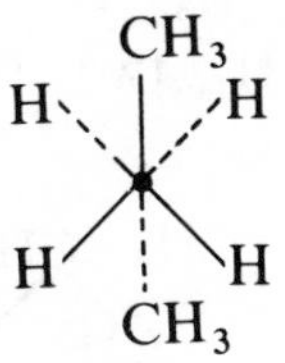

The latter staggered form keeps the two bigger methyl groups as far apart as possible and is the most stable. In this form the carbon atoms have a zig-zag arrangement.

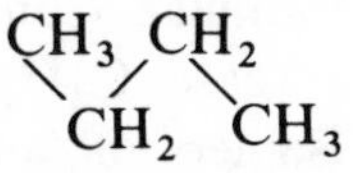

Such a zig-zag, staggered, shape is the preferred shape for all the alkenes, for example

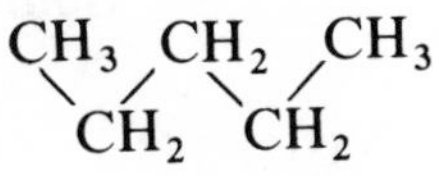

Pentane

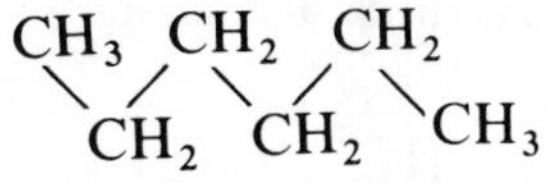

Hexane

The names methane, ethane, propane and butane are so-called trivial names in that they give no indication of the number of carbon atoms in these molecules. For molecules with more than four carbon atoms the length of the chain of carbon atoms is evident from the name; hence pentane and hexane, above, have five and six carbon atoms respectively in their chains of carbon atoms. Similarly we have C_7H_{16} [or $CH_3(CH_2)_5CH_3$] **heptane**, C_8H_{18} [or $CH_3(CH_2)_6CH_3$] **octane**, etc.

A series of compounds like this, with the same general formula, in this case C_nH_{n+2}, is known as a **homologous series.** All of its members are **homologues** of one another.

Branched-chain Alkanes

For alkanes with four or more carbon atoms further complications arise because it is possible to arrange the atoms in the form of branched chains as well as unbranched chains while still having a molecular formula C_nH_{2n+2}.

For C_4H_{10} two arrangements are possible, namely

$$CH_3 \diagdown CH_2 \diagup \diagdown CH_2 \diagdown CH_3 \qquad \text{and} \qquad CH_3 \diagdown CH \diagup CH_3 \quad | \quad CH_3$$

These are not different conformational forms of the same compound but rather two distinct compounds. They cannot be changed from one into the other without breaking chemical bonds. They are said to have different **structural formulae.** Compounds that have identical molecular formulae (in this case C_4H_{10}) but different structural formulae are described as **isomers** of one another. **Isomerism** is an all-pervading facet of organic chemistry.

Thus molecular formulae are commonly inadequate to represent an organic molecule and structural formulae must be used. Common shorthand versions are as shown for the two C_4H_{10} isomers:

$$CH_3 \diagdown CH_2 \diagup \diagdown CH_2 \diagdown CH_3 \qquad \qquad CH_3 \diagdown CH_3 \diagup \diagdown CH \quad | \quad CH_3$$

$$\text{or} \qquad \qquad \qquad \text{or}$$

$$CH_3(CH_2)_2CH_3$$
or

$$CH(CH_3)_3$$
or

Butane   2-Methylpropane*

In the first line the formulae are drawn to show the shape of the carbon skeleton of the molecules. In the second line, space-saving formulae are given, which do not show the shape but do show how the carbon atoms are attached to one another. [Note that $CH(CH_3)_3$ signifies a carbon atom with three methyl groups directly attached to it.] The linear formulae in the final line are space saving and time saving; they only show the carbon–carbon bonds in their correct geometry. Each carbon atom is taken to be attached to a number of hydrogen atoms sufficient to satisfy its covalency of four. The presence of any element other than carbon and hydrogen in the molecule must be specifically shown with an atomic symbol. Thus the following linear formulae would represent, respectively, $CH_3CH_2OCH_2CH_3$ and $CH_3CH_2CH_2Cl$.

and

All these three types of formulae are in common use, each type having its advantage for certain purposes.

Only the linear molecule is called butane. The branched molecule is called 2-methylpropane.[†] To obtain the name of a branched molecule, the longest chain of carbon atoms is taken as the root of the name and other groups attached to it are listed (in alphabetical order). Their place of attachment is signified by numbering the carbon atoms in the main chain (starting from the end of the chain). Thus 2-methylpentane is

$$\left[\text{or} \quad (CH_3)_2CHCH_2CH_2CH_3 \quad \text{or} \quad \right]$$

---

*In this instance the number 2- is not strictly necessary; if the extra methyl group were attached to the 1-carbon atom it would merely turn it into butane. Thus only one structure is possible for 'methylpropane'.

[†]Again the 2- is not strictly necessary; see above.

[This might also be called 4-methylpentane by numbering the chain from the other end, but the convention is to number from the end that provides the lowest number(s) for the substituent(s).] In the same way that the name of the group **methyl** CH_3— is derived from methane, CH_4, so the names of other substituent groups are derived from the related alkanes, for example CH_3CH_2—, **ethyl** from ethane.

For C_5H_{12}, three isomers are possible, namely:

$$CH_3 \overset{CH_2}{\diagdown} \overset{CH_2}{\diagup} \overset{}{\diagdown} CH_2 \diagup CH_2 \diagdown CH_3$$

$$\overset{CH_3}{\underset{|}{CH}} \overset{}{\diagup} \overset{}{\diagdown} CH_3 \qquad CH_3 \overset{CH_3}{\underset{\underset{CH_3}{|}}{\overset{|}{-C-}}} CH_3$$

or $CH_3(CH_2)_3CH_3$ or $(CH_3)_2CHCH_2CH_3$ or $(CH_3)_4C$

or   /\/\   or   \|/\   or   X

      Pentane   2-Methylbutane*   2,2-Dimethylpropane

The number of possible isomers rises rapidly with the number of carbon atoms in the alkane. This is shown in the following table:

Molecular formula	Number of possible isomers
C_4H_{10}	2
C_5H_{12}	3
$C_{10}H_{22}$	75
$C_{15}H_{32}$	4347
$C_{20}H_{42}$	366,319
$C_{40}H_{82}$	$\sim 62.5 \times 10^{12}$

Distinguishing Between Isomers of Alkanes

To distinguish between isomeric alkanes by chemical means is extremely difficult, because they all have very similar chemical

---

*Again the 2- is not strictly necessary; see above.

properties (see Chapter 6). Because isomers are different compounds they have different physical properties, e.g. melting points and boiling points, but these do not necessarily indicate what the structures of the molecules are.

Probably the simplest technique to use is that of *nuclear magnetic resonance spectroscopy* or *n.m.r.* A rather longer account of this technique is given in the Appendix. Put very simply, under appropriate conditions (in a magnetic field) a sample of material can absorb radiation in the radio-frequency region of the spectrum, that is $\sim 10^7$ hertz. This absorption is associated with the magnetic moments of the nuclear particles. In general, nuclei with an even mass number and an even atomic number, for example ^{12}C, do not give rise to n.m.r. spectra, but nuclei which have an odd mass number or atomic number do provide n.m.r. spectra. The most important of these for organic chemistry are ^1H and ^{13}C, the latter being a minor constituent of all natural carbon. ^1H and ^{13}C absorptions come in slightly different parts of the spectrum and, for operational reasons, are recorded separately.

The absorption connected with any atom depends upon the chemical environment of that atom. Hence it is possible to distinguish between atoms of the same elements which are in different chemical environments.

Consider ethane, CH_3—CH_3. Both carbon atoms occupy identical chemical environments, i.e. each is surrounded by three hydrogen atoms and a methyl group. Similarly all the hydrogen atoms are in identical environments, each being attached to an ethyl group —CH_2CH_3. Hence both the ^1H-n.m.r. spectrum and the ^{13}C-n.m.r. spectrum of ethane show only one absorption signal.

In the case of propane, $CH_3CH_2CH_3$, there are two different sorts of carbon atom, namely those forming part of the two methyl groups at the ends of the molecule and the one in the middle of the molecule, which is part of a **methylene group** (—CH_2—). Similarly, there are two different sorts of hydrogen atoms, six associated with the methyl groups and two with the central methylene group. Therefore in this case, two signals appear in the ^{13}C-n.m.r. spectrum and in the ^1H-n.m.r. spectrum of propane. In the case of the ^{13}C-n.m.r. spectrum, under normal operating conditions, the two signals appear as two sharp lines, but in the case of the ^1H spectrum, the signals appear as somewhat broader signals and furthermore have a structured appearance, being made up of a number of adjacent peaks. In the case of ^1H-n.m.r. spectra, but **not** normally of ^{13}C-n.m.r. spectra, the ratio of the sizes of the peaks (actually their areas), which is recorded and displayed on the spectrum, shows the ratio of the numbers of hydrogen atoms of each sort. Thus in the case of propane the ratio of the sizes of the peaks is 2:6.

Further information about n.m.r. spectroscopy is given in the Appendix. Some typical examples are shown on the facing page.

As a final example at this stage, let us apply this method to distinguish between the isomers of molecular formula C_5H_{12}.

(i)	(ii)	(iii)
^{1}H-n.m.r.: 3 signals (2:4:6)	4 signals (1:2:3:6)	1 signal
^{13}C-n.m.r.: 3 signals	4 signals	2 signals

In the case of (iii) all the hydrogen atoms are identical; its structural formula can be written as $C(CH_3)_4$. It has two sorts of carbon atom: those (a) that are part of methyl groups and that (b) to which the four methyl groups are attached.

Isomer (i) has three different kinds of carbon atoms (a, b, c) and three different kinds of hydrogen atoms (a, b, c). The methyl groups at the ends of the chain have one sort of carbon and hydrogen (a). The three methylene (CH_2) groups are **not** all the same. The central one (c) is directly attached to two other methylene groups (b), but the other methylene groups (b) are each directly attached to a methyl group (a) and a methylene group (c). Hence isomer (i) provides three signals in both its ^{1}H- and its ^{13}C-n.m.r. spectrum, and the ^{1}H signals are in the ratio 2:4:6.

Isomer (ii) provides four signals in both its ^{1}H-n.m.r. and ^{13}C-n.m.r. spectra. The carbon and hydrogen atoms forming part of the CH group (b) and CH_2 group (c) are evidently distinct. Of the three methyl groups, those marked (a) are identical, both being linked to the CH group, but that marked (d) is in a different environment, being attached to the CH_2 group. The ratio of the sizes of the ^{1}H-n.m.r. signals is 1(b):2(c):3(d):6(a).

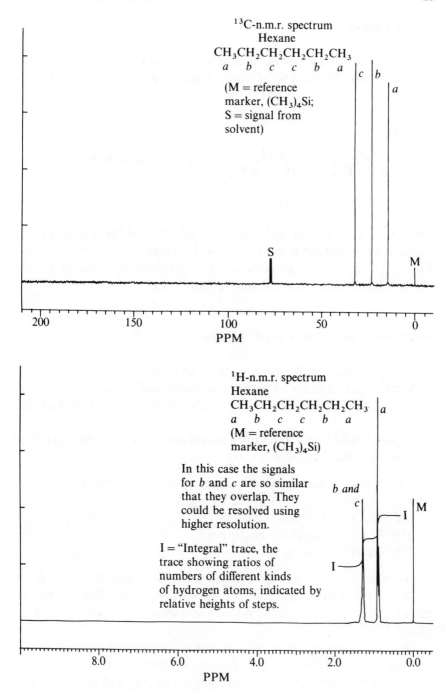

¹³C-n.m.r. spectrum
Hexane
$CH_3CH_2CH_2CH_2CH_2CH_3$
a   b   c   c   b   a

(M = reference
marker, $(CH_3)_4Si$;
S = signal from
solvent)

S

M

c b

a

200   150   100   50   0
PPM

¹H-n.m.r. spectrum
Hexane
$CH_3CH_2CH_2CH_2CH_2CH_3$
a   b   c   c   b   a
(M = reference
marker, $(CH_3)_4Si$;

In this case the signals
for *b* and *c* are so similar
that they overlap. They
could be resolved using
higher resolution.

I = "Integral" trace, the
trace showing ratios of
numbers of different kinds
of hydrogen atoms, indicated by
relative heights of steps.

a

b and
c

I

M

I

8.0   6.0   4.0   2.0   0.0
PPM

Examples of n.m.r. spectra

In Conclusion

In conclusion, alkanes have the general formula C_nH_{2n+2}. Their names all end in -**ane**. An example of a branched-chain alkane with its name is

$$
\begin{array}{c}
CH_3 \\
\quad CH_2 \\
CH_2 \quad CH \quad CH_2 \\
CH_3 \quad CH_2 \quad CH \quad CH_3 \\
\qquad\qquad CH_3
\end{array}
$$
4-Ethyl-3-methylheptane

Groups such as methyl and ethyl, formally derived from an alkane by loss of a hydrogen atom, are called **alkyl** groups. They have the general formula C_nH_{2n+1}. Alkyl groups are commonly represented by the symbol R; alkanes are then represented as RH.

Sources and Everyday Uses of Alkanes

The great source of alkanes is natural gas and oil.

Natural gas consists largely of methane and is used as a fuel in Britain, where it is obtained from the North Sea, and throughout the world.

Crude oil is first distilled. **Very approximately** the fractions that distil may be described as follows:

b.p. $< 30\,°C$	gaseous fraction (gas cylinders, e.g. butane)
$30-180\,°C$	petrol (in the USA = gasoline)
$180-230\,°C$	kerosene (jet fuel)
$230-300\,°C$	diesel oil, heating oil
$300-400\,°C$	heating oil

There remains a large amount of involatile material. Among the products obtained from this are lubricating oils, paraffin waxes, petroleum jelly and asphalt. All of these fractions consist largely of alkanes.

The two most important uses of these alkanes are as fuels and as starting materials from which an enormous range of synthetic products are obtained. There is more about the preparation and importance of some of these products in later chapters, and also about the use of alkanes as fuels in Chapter 6, wherein the chemical properties of alkanes are discussed.

The alkanes with higher boiling points have higher molecular weights; this can be illustrated by the following diagram:

$$\overrightarrow{\frac{\text{Increasing molecular weight}}{\text{Gases} \quad \text{Liquids} \quad \text{Greases} \quad \text{Waxes}}}$$

Crude oil does not contain enough constituents of sufficiently low molecular weight to provide all the petrol and oil required. If the higher boiling material could be converted into smaller alkanes, this provides a bigger supply of alkanes suitable for use as petrol and oil. This involves the breaking of carbon chains in the bigger alkanes to provide smaller molecules and can indeed be achieved with the help of suitable catalysts. The process is known as **cracking** of hydrocarbons.

Physical Properties of Alkanes

The alkanes of low molecular weight are volatile; those of high molecular weight are greasy. All are insoluble in water. Why do they have these physical characteristics?

Alkane molecules are made up of atoms of carbon and of hydrogen. Atoms of both of these elements are of similar **electronegativity**, i.e. they both have a similar affinity for electrons. In consequence the electrons in the bonds that hold the atoms together in alkane molecules are rather evenly shared between the atoms they link. Hence the bonds are not **polarized**; in other words the electrons of the bond do not tend to be associated more with one of the atoms than with the other as happens, for example, in the case of carbon–chlorine bonds in which the electrons tend to be associated more closely with the chlorine atom than with the carbon atom. The chlorine atom is of higher electronegativity than the carbon atom. This can be represented schematically by an arrowhead drawn on the bond linking the atoms, viz.

$$C \longrightarrow Cl$$

The arrow points towards the atom more associated with the bonding electrons. In the extreme the electrons may be almost entirely associated with one element, as in sodium chloride, where a crystal is made up from alternate positively charged sodium ions and negatively charged chloride ions:

$$Na^+Cl^- \quad Na^+Cl^- \quad \text{etc. (in all dimensions)}$$

Because there is little polarization in alkanes there are only very weak intermolecular forces in these compounds. Hence the molecules do not interact strongly with one another. This in turn leads to the alkanes having low melting points and boiling points. Lacking polar interactions, it is also possible for the molecules to slide past each other easily, resulting in their greasy texture. There is very little intermolecular interaction between the non-polar alkanes and highly polar molecules such as water—hence the insolubility of alkanes in water and the immiscibility of oil and water.

Cycloalkanes

In oil there are molecules which very closely resemble alkanes but which have cyclic structures, for example

or

Cyclopentane

Because of the cyclic structure these molecules have molecular formulae with two hydrogen atoms less than an uncyclized alkane with the same number of carbon atoms. Thus the molecular formula of cyclopentane is C_5H_{10}. Note that the ring is completely symmetrical and all the CH_2 groups are equivalent. Hence this compound will show only one signal in both its ^1H-n.m.r. and ^{13}C-n.m.r. spectra. Any size of ring is possible. These cyclic alkanes are known as **cycloalkanes**. Their names indicate the number of carbon atoms in the ring with the prefix **cyclo** added. Thus C_3H_6 is cyclopropane and C_6H_{12} is cyclohexane. Alkyl-substituted cycloalkanes also exist, for example

or

Methylcyclopentane

The shape of these rings is important. Although it is in fact a little bent, the cyclopentane ring has a more or less planar shape:

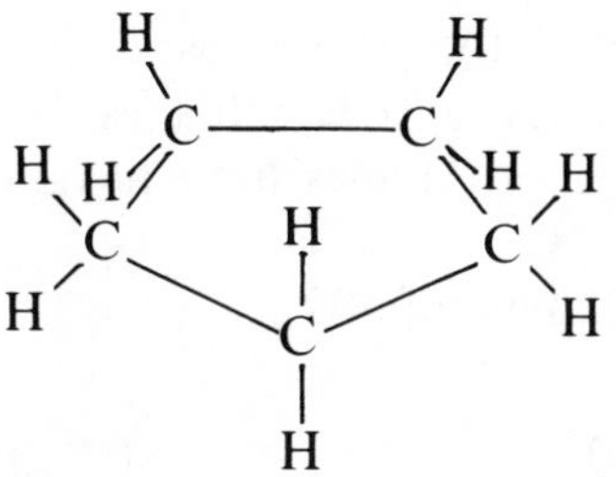

In consequence all the CH_2 groups have eclipsed conformations with respect to one another.

In contrast the cyclohexane ring is described as 'chair shaped':

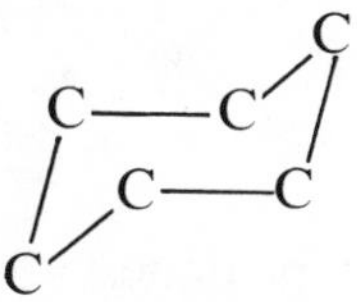

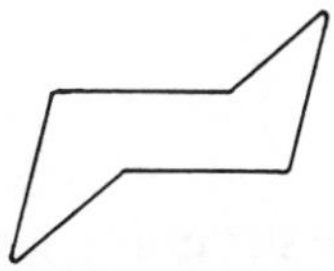

The ring takes up this shape because in this way the carbon atoms can take up their preferred tetrahedral shapes; in a planar ring the C—C angles would be too big. In this chair shape the hydrogen atoms are all in staggered conformations:

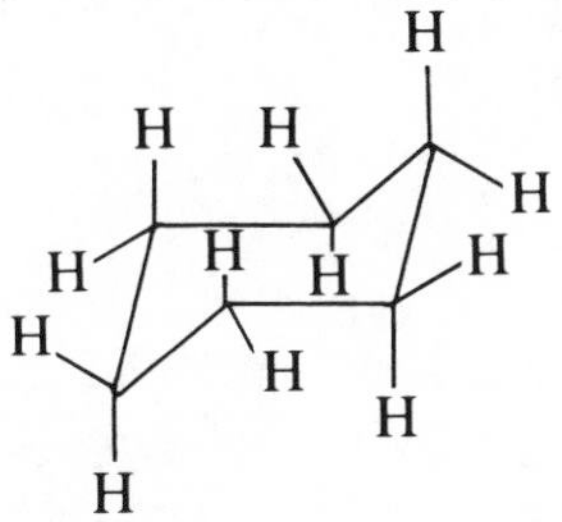

or, viewed along one of the carbon–carbon bonds:

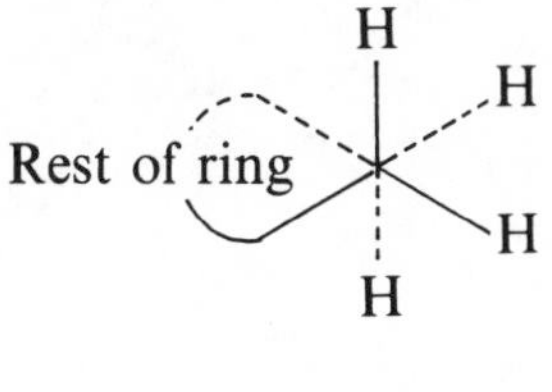

It is again strongly recommended that the reader makes models, which will show the shapes of these molecules much more clearly than diagrams.

Cyclobutane and especially cyclopropane molecules are rather less stable than other alkanes or cycloalkanes because inevitably, to provide the rings, the C$\diagdown$C angles have to be distorted away from the normal tetrahedral angle of 109°:

Such molecules are said to be **strained** and this affects their properties.

Questions

1. Write the structural formulae of 3-methylpentane, cyclohexane, 1, 3-dimethylcyclopentane.
2. Write the structural formulae and names of all the isomers that have molecular formulae (a) C_5H_{12} and (b) C_6H_{14}.
3. Draw the preferred conformations of propane and of butane.
4. How many signals would be observed in the ^{1}H-n.m.r. and ^{13}C-n.m.r. spectra of (a) butane, (b) 2-methylbutane, (c) 2-methylpentane, (d) 2, 3, 4-trimethylpentane, (e) 2, 2-dimethylpropane and (f) cyclopentane?

3 ALKENES AND THEIR STRUCTURE

In the last paragraphs of the previous chapter mention was made of cycloalkanes, which have the general molecular formula C_nH_{2n}.

There is also another series of compounds, called **alkenes**, with this general formula C_nH_{2n}. Thus there is a compound C_3H_6, **propene**, and also C_2H_4, **ethylene**, which has no cycloalkane equivalent.

If carbon is to retain its covalency of four and hydrogen its covalency of one, the only way to draw a formula for ethylene is with **two** bonds linking the two carbon atoms, viz.

$$\begin{array}{ccc} H\diagdown & & \diagup H \\ & C{=}C & \\ H\diagup & & \diagdown H \end{array} \quad \text{or} \quad CH_2{=}CH_2$$

This is described as a **double bond**. Similarly, propene may be represented as

$$\begin{array}{ccc} H\diagdown & & \diagup H \\ & C{=}C & \\ H\diagup & & \diagdown CH_3 \end{array} \quad \text{or} \quad CH_2{=}CHCH_3$$

Note that n.m.r. spectroscopy would readily distinguish between propene and cyclopropane for the latter compound provides only one signal in its ^{13}C-n.m.r. spectrum, whereas propene has three different sorts of carbon atoms, namely those associated with the CH_2, CH and CH_3 groups respectively.

As with alkanes, the alkenes of low molecular weight are gases at normal room temperature. Like alkanes, as the molecular weights increase, they are liquids or, for the larger molecules, solids; the C_5H_{10} isomers have boiling points in the range 20–40 °C.

Nomenclature of Alkenes

Alkenes are named in the same way as alkanes, save that the names end in -*ene* rather than -*ane*. Hence the name *propene*. The systematic name for ethylene is *ethene* but the non-systematic name *ethylene* is the one more commonly met. It is frequently the case that common simple organic molecules are known best by their so-called trivial names, and these names must be learned since they are in everyday usage. This usage of trivial names for some simple organic compounds is in accord with the international rules for nomenclature.*

When we come to C_4H_8 two isomers are possible which differ only in the placement of the double bond; there is a third isomer having a branched chain:

$$CH_2{=}CHCH_2CH_3 \qquad CH_3CH{=}CHCH_3 \qquad (CH_3)_2C{=}CH_2$$
But-1-ene   But-2-ene   Methylpropene

The first two isomers are distinguished in their names by including a number which shows where the double bond is placed. In but-1-ene it lies between carbon atoms 1 and 2; in but-2-ene it lies between carbon atoms 2 and 3. Note that once again the lowest possible number is used to define the position.[†]

No numbering is needed for methylpropene because no other isomer is possible involving this arrangement of carbon atoms.

For other branched-chain alkenes numbering may be needed, for example

$$\underset{\text{2-Methylbut-1-ene}}{CH_2{=}\overset{\overset{\textstyle CH_3}{|}}{C}{-}CH_2CH_3} \qquad \underset{\text{3-Methylbut-1-ene}}{CH_2{=}CH{-}CH(CH_3)_2 \;(\text{or } CH_2{=}CHCHCH_3)} \qquad \underset{\text{2-Methylbut-2-ene}}{CH_3\overset{\overset{\textstyle CH_3}{|}}{C}{=}\overset{\overset{\textstyle CH_3}{|}}{C}HCH_3}$$

---

*The necessity to know trivial names in common everyday usage finds a parallel in the use of human names. Bill's parents may have called him William, Betty's parents may have named her Elizabeth, but if they are called Bill and Betty in everyday life, it is vital to recognize those names. A comment from a recent (1987) International Union of Pure and Applied Chemistry publication is relevant: 'The use of trivial names has deep historical roots. It makes communication among chemists easier and thus can hardly be terminated by a decree for a more rigorous but less customary nomenclature.'

[†] In the United States these names become 1-butene and 2-butene. The simpler British system always puts the numeral **directly** before the group to which it refers.

Note that the lowest possible number is always given to indicate the position of the double bond.

Molecules with more than one double bond exist and are named similarly, but the number of double bonds must be indicated by describing them as diene, triene, tetraene, etc. Examples are

$$CH_2=CH-CH_2-CH=CH_2$$
Penta-1,4-diene

$$CH_3-CH=CH-CH=CH-CH=CH_2$$
Hepta-1,3,5-triene

Geometry of Alkenes

Let us first consider ethylene:

$$\begin{array}{c} H \\ \diagdown \\ H \diagup \end{array} C=C \begin{array}{c} H \\ \diagup \\ \diagdown H \end{array}$$

If the bonds linking the atoms are to be as far apart from one another as possible, as mentioned in the previous chapter (page 15), the molecule will take up a flat structure and the bonds attached to each carbon atom will be at angles of 120° to one another. Such carbon atoms are called **trigonal**, in contrast to the tetrahedral carbon atoms in alkanes. This structure for alkenes is confirmed by experimental results.

The double bond is not identical to two single bonds.* Thus it is stronger than a single bond ($\sim 610\,kJ$) but less strong than two single bonds ($2 \times \sim 350\,kJ = 700\,kJ$). (These numbers represent the mean energies required to break such bonds; cf. Chapter 1.)

A carbon–carbon double bond is shorter ($\sim 1.34\,\text{Å}$) than a carbon–carbon single bond ($\sim 1.54\,\text{Å}$), i.e. the two carbon atoms joined by the double bond are closer together than those in an alkane.

One result of the tighter bonding in a double bond is that rotation about a double bond requires much more expenditure of energy than does rotation about a single bond. Indeed such rotation requires a rather larger input of energy, $\sim 170\,kJ$, than is readily available at room temperature. In consequence rotation about a double bond does not commonly take place at 'normal' temperatures. As a result of this there

---

*A discussion of the nature of the bonding is not necessary at this stage to 'explain' the properties of alkenes and will therefore not be included.

are two 'sides' to a double bond. If there are two different groups, X, Y, attached at each end of a double bond, two isomers are possible depending on which 'side' of the double bond they are located:

Because rotation about the double bond is inhibited these are different compounds. They are often described as *cis* and *trans* isomers, *cis* indicating that two similar groups are on the *same* side of the double bond, *trans* that they are on opposite sides. Alternatively they are called Z and E isomers, this nomenclature deriving from the German words *zusammen* (= together = Z) and *entgegen* (= opposite = E). Specific examples are:

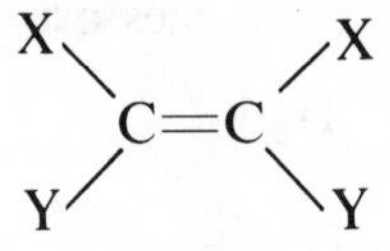

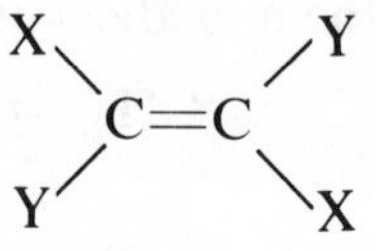

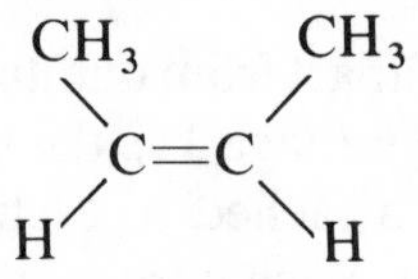

cis-But-2-ene
or   Z-but-2-ene

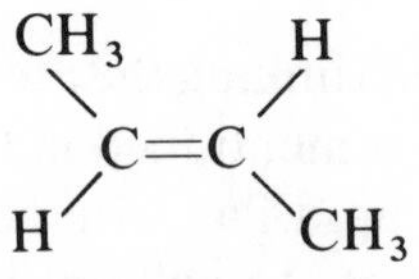

trans-But-2-ene
or   E-but-2-ene

E and Z isomers frequently have quite different chemical and physical properties.

It may be noted that a similar sort of isomerism is possible for cycloalkane derivatives, for example

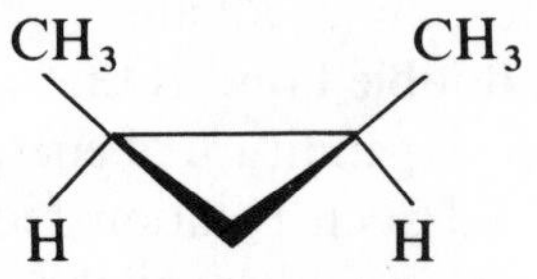

cis-1, 2-Dimethyl-
cyclopropane

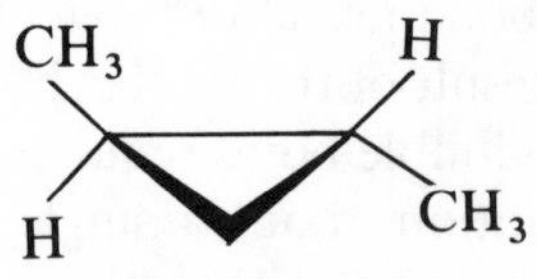

trans-1, 2-dimethyl-
cyclopropane

This sort of isomerism, in which isomers differ only by being *cis* or *trans* forms, is called **geometric isomerism**.

Questions

1. How many geometric isomers are there of (a) but-2-ene, (b) but-1-
   ene, (c) hexa-2,4-diene and (d) 1,3-dimethylcyclopentane?
2. How many signals would be observed in ^{1}H-n.m.r. and ^{13}C-n.m.r.
   spectra of (a) ethylene, (b) E-but-2-ene and (c) Z-but-2-ene?
3. Write structural formulae for all the isomers of molecular formula
   C_4H_8.

4 THE ROLE OF ENERGY IN CHEMICAL REACTIONS

Alkenes have two 'extra' electrons more than the bare number required to hold all the atoms together, and these electrons can take part in reactions with other atoms. For example, ethylene reacts with bromine:

$$CH_2{=}CH_2 + Br_2 \longrightarrow \overset{\overset{\displaystyle Br}{\displaystyle |}}{CH_2}{-}\overset{\overset{\displaystyle Br}{\displaystyle |}}{CH_2}$$

Since the net result is the addition of one molecule of bromine to one molecule of ethylene this is described as an **addition reaction**.

This reaction involves not only the formation of new bonds linking the bromine atoms to the carbon atoms but also the breaking of the bond that initially joins the two bromine atoms together.

Formation of chemical bonds releases energy, and conversely the breaking of chemical bonds requires an input of energy.

If in the course of a chemical reaction more energy is required for the breaking of bonds than is released in the formation of new bonds, then extra energy must be supplied to enable the reaction to proceed. Such a process is described as **endothermic**.

If, however, the energy released in the formation of new bonds in a reaction exceeds the amount of energy required for the breaking of bonds in that reaction, then once the reaction starts, the overall result is a release of energy, usually as heat. Such reactions are called **exothermic**.

At first sight it might appear that the more exothermic the reaction the more easily it would go. However this is a rather naive over-simplification.

A few lines above it was stated that a reaction would be exothermic 'once the reaction starts'. The combustion of petrol is highly exothermic, but it needs the application of energy to start it, e.g. by applying a lighted match.

For a reaction between two molecules to take place a number of conditions must be fulfilled, viz.:

1. The reacting molecules must come close enough to one another to interact.
2. The molecules must be correctly aligned with one another. For example, in a very common type of reaction represented by

$$X—Y + Z \longrightarrow X + Y—Z$$

   it is necessary that Z approaches X—Y from the side of Y away from X.
3. The reacting molecules must have sufficient energy to overcome the repulsive forces which normally exist between different molecules.*

The progress of a chemical reaction may be expressed graphically in what is called a **reaction profile.** In these profiles the energy of the reacting molecules is plotted against the course of the reaction, i.e. time.

An energy profile for the reaction $X—Y + Z \rightarrow X + Y—Z$ might well be as follows:

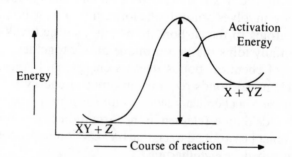

The energy input required to start the reaction is represented by the peak in the graph; the height of this peak is called the **activation energy** of the reaction. The state to which the reaction has proceeded at this point, i.e. the point of highest energy, is called the **transition state** of the

---

*These conditions have their analogies in cricket. To hit the ball a batsman must (1) ensure that his bat comes near the ball, (2) hold his bat in the correct position (at the simplest level, the right way round!) to meet the ball and (3) use enough energy to drive the ball to its desired destination.

reaction. This may represent an intermediate stage in the reaction, for example

$$X—Y + Z \rightarrow X \cdots Y \cdots Z \rightarrow X + YZ$$

when the X—Y bond is partially broken and the new Y—Z bond is partially formed.

In the energy profile drawn here the overall energy of the products at the end of the reaction is greater than the energy of the initial reactants, i.e. the reaction must be **endothermic** and require an overall input of energy for it to occur.

The energy profile for an **exothermic** reaction would be as follows:

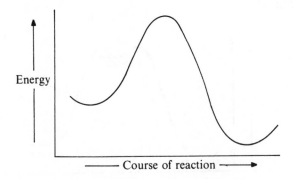

In each case, whether the reaction is endothermic or exothermic, the difference between the energies of the reactants and of the products is known as the **heat of reaction**.

When two molecules 'collide' their kinetic energy is partially converted into potential energy. If this energy is sufficient to attain the energy of the transition state, reaction takes place.

Reactions are thus often assisted by increasing the kinetic energy available. This is achieved most commonly by heating the reactants, or in some instances by irradiating them with light.

Any extraneous factor that can lower the activation energy of a reaction makes the reaction take place more readily. This may sometimes be achieved by the use of a catalyst (see diagram overleaf), and sometimes by the use of an appropriate solvent.

In the case of some reactions there is a halfway house between the reactants and products. Thus in the reaction X—Y + Z → X + Y—Z the reaction might proceed via an intermediate compound X—Y—Z, which then decomposes again to provide X and Y—Z. In such cases

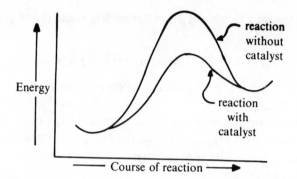

a **reactive intermediate** is said to be formed and the reaction profile has the form:

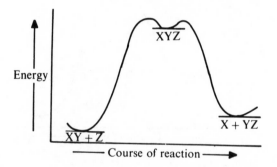

Note that each step in such a reaction, the formation of the reactive intermediate and its subsequent decomposition to provide the final products, has its own transition state, possibly

$$X\text{—}Y + Z \rightarrow X\text{—}Y\cdots Z \rightarrow X\text{—}Y\text{—}Z \rightarrow$$
$$X\cdots Y\text{—}Z \rightarrow X + Y\text{—}Z$$

Note that in the reactions we have just been considering, $X\text{—}Y + Z \rightarrow X + Y\text{—}Z$, by whatsoever mechanism the reaction takes place, whether it involves formation of a reactive intermediate or proceeds directly via a transition state, the net result is the same, namely the substitution of Z for X as the atom or group of atoms attached to Y. An example is

$$HO^- + CH_3Br \rightarrow HOCH_3 + Br^-$$

Such a reaction is therefore called a **substitution reaction** and the end result is that one atom or group of atoms is replaced by another atom or group of atoms. Thus in the above example a bromine atom is replaced by a hydroxy (HO) group.

Question

1. Draw the energy profiles for (a) a simple endothermic reaction, (b) a simple exothermic reaction and (c) an exothermic reaction in which a reactive intermediate is formed between the initial and final stages of the reaction.

5 REACTIONS OF ALKENES

It was mentioned at the beginning of the previous chapter that ethylene undergoes an **addition reaction** with bromine to give 1,2-dibromoethane:

$$CH_2{=}CH_2 \xrightarrow{\text{Br}_2} BrCH_2{-}CH_2Br$$

Addition reactions are characteristic reactions of alkenes:

$$R_2C{=}CR_2 \xrightarrow{X{-}Y} XCR_2{-}CR_2Y$$

Because they readily add other atoms alkenes are described as **unsaturated** compounds.

Reactions with Hydrogen Halides

Consider as an example the reaction of ethylene with hydrogen chloride:

$$CH_2{=}CH_2 + HCl \rightarrow CH_3CH_2Cl$$

The electrons that bond the two atoms together in hydrogen chloride are not associated to the same extent with both atoms. Chlorine is of higher electronegativity than hydrogen and the bonding electrons are more associated with the chlorine atom. Hence the H—Cl bond is polarized and may be represented, as discussed in Chapter 2, by the symbol H $\rightarrow$ Cl.

Since the reaction with ethylene involves the 'spare' electrons of this molecule (see the beginning of Chapter 4), it is a reasonable guess that the reaction between an ethylene molecule and a molecule of hydrogen chloride may start by interaction of these 'spare' electrons with the electron-deficient hydrogen atom of hydrogen chloride. This may be depicted pictorially as follows:

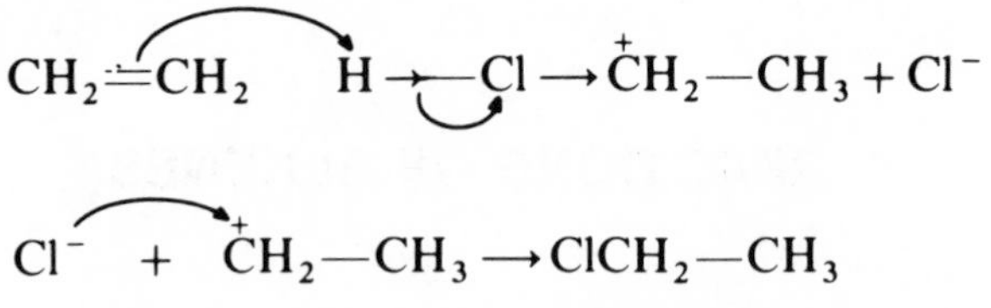

$$CH_2 \!=\! CH_2 \quad H \!-\! Cl \rightarrow \overset{+}{C}H_2 \!-\! CH_3 + Cl^-$$

$$Cl^- \;+\; \overset{+}{C}H_2 \!-\! CH_3 \rightarrow ClCH_2 \!-\! CH_3$$

In this representation the curved arrows denote the movement of a pair of electrons. The tail of the arrow indicates the source of the pair of electrons and the head of the arrow indicates the site to which they move. This representation is quite precise, and arrows must not be drawn other than with this meaning. For example, to depict the reaction of ethylene and hydrogen chloride as

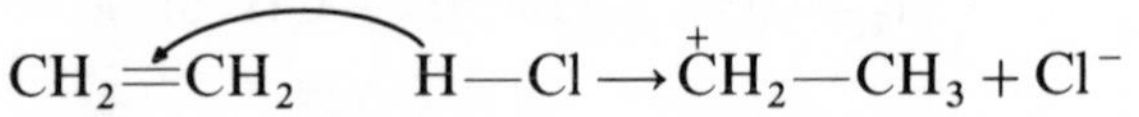

$$CH_2 \!=\! CH_2 \quad H \!-\! Cl \rightarrow \overset{+}{C}H_2 \!-\! CH_3 + Cl^-$$

is completely **wrong**: the electron-deficient hydrogen atom **cannot** be a source of electrons.

The two electrons which made up the hydrogen chloride bond become associated with the chlorine atom, thus providing a chloride anion.

The two electrons from the double bond attach the hydrogen atom to one of the carbon atoms. The other carbon atom is now left with only six electrons associated with it; it is electron–deficient and consequently acquires a positive charge. A positively charged carbon atom is described rather confusingly, sometimes as a **carbenium ion**, sometimes as a **carbonium ion** and sometimes as a **carbocation**; all three of these names may be encountered in different writings. Carbenium ion is probably the better term, and is used in this text. In the second stage of the addition reaction this positively charged carbon atom reacts with a negatively charged chloride ion to produce a carbon–chlorine bond (see the reaction scheme above), resulting in the formation of chloroethane (also called, less satisfactorily, ethyl chloride) as the product of the addition reaction.

From the mechanism proposed for the reaction of ethylene with hydrogen chloride, it could be expected that other acids might react similarly, and indeed the following reactions take place:

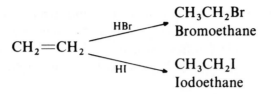

A similar reaction also occurs with 98% sulphuric acid, providing monoethyl sulphate:*

$$CH_2=CH_2 \xrightarrow{98\% \, H_2SO_4} CH_3CH_2OSO_3H \; (\text{or } CH_3CH_2OSO_2OH)$$

On treatment with water this compound is converted into ethanol and sulphuric acid:

$$CH_3CH_2OSO_3H + H_2O \longrightarrow CH_3CH_2OH + HOSO_3H$$

With aqueous solutions of sulphuric acid ethylene gives ethanol directly. This happens because the carbenium ion which is formed initially reacts with water preferentially to reacting with the hydrogen sulphate anion:

$$CH_2=CH_2 \; + \; H-OSO_3H \longrightarrow \overset{+}{C}H_2-CH_3 + \; ^-OSO_3H$$

$$H_2\overset{..}{O}: \; \overset{+}{C}H_2CH_3 \; \longrightarrow \; \underset{H}{\overset{H}{\diagdown}}\overset{+}{O}-CH_2CH_3 \longrightarrow HOCH_2CH_3$$

Reaction involves the lone pair of electrons associated with the oxygen atom of water. Addition of water to the carbenium ion provides an **oxonium ion**, having a tervalent positively charged oxygen atom, which readily loses a proton to give ethanol.

Two reasons conspire to make water react rather than the hydrogen sulphate anion. The latter anion is very stable and unreactive in any case, since it is an anion derived from a strong acid (see Chapter 7), and, in addition, in aqueous solution, far more water molecules than hydrogen sulphate anions are available and so their reaction is also statistically favoured.

The conversion of ethylene into ethanol is an extremely important

*Monoethyl sulphate signifies that in this compound one of the hydrogen atoms of sulphuric acid ($HOSO_2OH$) is replaced by an ethyl group. Similarly $C_2H_5OSO_2OC_2H_5$ is diethyl sulphate.

industrial process, since there is a large industrial requirement for ethanol, and ethylene is a very readily available starting material (derived from crude oil) from which to obtain it. On the industrial scale ethylene is treated with aqueous solutions of either sulphuric acid or phosphoric acid.

Electrophiles and Nucleophiles

Reagents that react with excess electrons in a molecule, as happens in the above addition reactions of acids with ethylene, are called **electrophiles** (i.e. 'electron lovers'; cf. 'bibliophiles' for lovers of books). In general a reagent that reacts with electron-rich centres in molecules is called an electrophile; the reaction is described as **electrophilic** attack on the molecule. Alkenes are electron-rich and undergo electrophilic attack by acids HX. Reagents having negative charge, for example Cl^-, or lone pairs of electrons, as in water, attack positively charged centres, or in more general terms electron-poor centres, in molecules and are called **nucleophiles**. (Atomic nuclei are positively charged; this provides the source of this name.) Thus in the formation of ethanol from ethylene, the intermediate carbenium ion undergoes **nucleophilic** attack by water.

Reactions with Halogens

Alkenes undergo addition reactions with halogens, for example

$$CH_2{=}CH_2 \underset{Br_2}{\overset{Cl_2}{\diagup\diagdown}} \begin{array}{l} ClCH_2CH_2Cl \quad 1,2\text{-Dichloroethane} \\ \\ BrCH_2CH_2Br \quad 1,2\text{-Dibromoethane} \end{array}$$

These reactions are normally carried out using an organic solvent.

How do these reactions occur since the Br–Br and Cl–Cl bonds are not dipolar? Usually there are dipolar molecules in the immediate environment of the reaction, provided by the presence of moisture (water) or of the glass surface of the reaction vessel. These dipolar surroundings can induce polarization of the alkene and/or halogen molecules (much as magnetism can be induced by the presence of a magnet), and in this way reaction can be initiated, for example

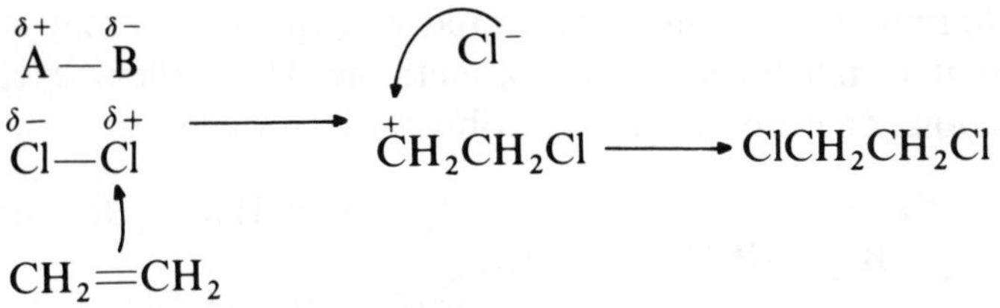

$$\overset{\delta+}{A}\!-\!\overset{\delta-}{B}$$

$$\overset{\delta-}{Cl}\!-\!\overset{\delta+}{Cl}$$

$$CH_2\!=\!CH_2$$

$$\overset{+}{C}H_2CH_2Cl \longrightarrow ClCH_2CH_2Cl$$

$$Cl^-$$

where A—B is some dipolar molecule in the vicinity of the reactants. A similar picture may be given for the reaction involving bromine.

In support of this mechanism, if a mixture of the alkene and halogen are kept together under rigorously pure conditions away from dipolar molecules in a flask with a waxed surface, little reaction between them ensues.

Further evidence for this mechanism, and that the reaction does indeed go in two steps, is provided by the mixture of products obtained if ethylene is passed into a solution of bromine in aqueous sodium chloride. A mixture of three products is obtained:

$$CH_2\!=\!CH_2 + Br_2 + H_2O + NaCl \longrightarrow \begin{cases} BrCH_2CH_2Br \\ BrCH_2CH_2Cl \\ BrCH_2CH_2OH \end{cases}$$

Neither chloride ions nor water react with ethylene in the absence of bromine, so the first step here must be a reaction of ethylene with bromine to provide a species that will then react with water or chloride ions in a second step.

Since chloride ions are negatively charged and since, as we have seen above, water is also a nucleophile, it must be assumed that the intermediate with which they react bears a positive charge, i.e. that the overall reactions are as follows:

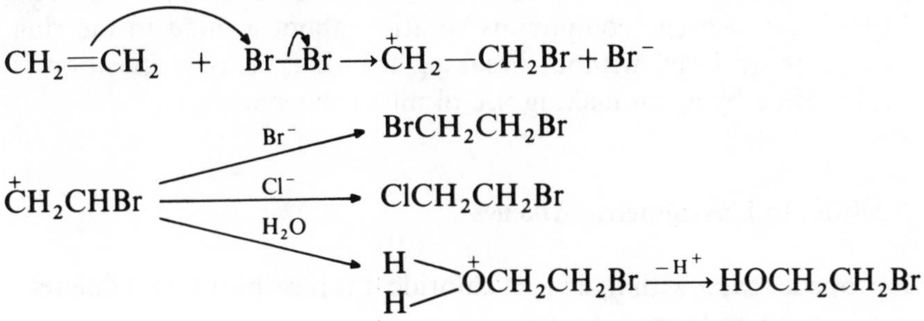

$$CH_2\!=\!CH_2 + Br\!-\!Br \longrightarrow \overset{+}{C}H_2\!-\!CH_2Br + Br^-$$

$$\overset{+}{C}H_2CHBr \quad \begin{cases} \xrightarrow{Br^-} BrCH_2CH_2Br \\ \xrightarrow{Cl^-} ClCH_2CH_2Br \\ \xrightarrow{H_2O} \overset{H}{\underset{H}{>}}\overset{+}{O}CH_2CH_2Br \xrightarrow{-H^+} HOCH_2CH_2Br \end{cases}$$

An alternative possibility that could be suggested is that ethylene and bromine first form 1,2-dibromoethane which then reacts with chloride ions or water to give the observed products:

$$CH_2{=}CH_2 + Br_2 \longrightarrow BrCH_2CH_2Br \underset{H_2O}{\overset{Cl^-}{\underset{\longrightarrow}{\overbrace{\hspace{2cm}}}}} \begin{matrix} ClCH_2CH_2Br + Br^- \\ HOCH_2CH_2Br + H^+ + Br^- \end{matrix}$$

Were this to be the case it could be expected that **both** bromine atoms of the dibromoethane might be replaced by chloride, giving 1,2-dichloroethane, but this does not happen.

When ethylene reacts with aqueous solutions of chlorine or bromine the major products are 2-chloroethanol, $ClCH_2CH_2OH$, or 2-bromoethanol, $BrCH_2CH_2OH$, respectively. Other alkenes react similarly.

If a cyclic alkene reacts with bromine the resultant addition product might have the bromine atoms both on the same side of the ring, *cis* to each other, or on opposite sides of the ring, *trans* to each other:

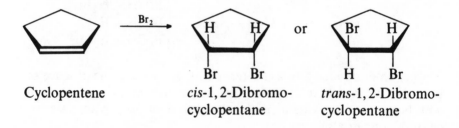

| Cyclopentene | *cis*-1,2-Dibromo-cyclopentane | *trans*-1,2-Dibromo-cyclopentane |

In fact the product is almost entirely the *trans* isomer. No such geometric isomerism is possible in the products derived from addition of bromine to unbranched non-cyclic alkenes because the open-chain products can readily be rotated about the single bond (see Chapter 2). In the case of cyclic compounds rotation about a bond in the ring cannot be achieved without breaking the ring. (This proposition is easily tested by again making use of molecular models.)

Addition to Unsymmetric Alkenes

If propene reacts with hydrogen chloride it is possible for two different products to be formed, that is

$$CH_3—CH=CH_2 + HCl \begin{cases} CH_3CHClCH_3 & \text{2-Chloropropane} \\ CH_3CH_2CH_2Cl & \text{1-Chloropropane} \end{cases}$$

In fact the major product is 2-chloropropane.

Note that it is easy to distinguish between these compounds, especially by their ^{13}C-n.m.r. spectra. 2-Chloropropane has **two** different kinds of carbon atom, namely the two methyl groups which are identical and the central carbon atom; hence it will give two signals in its spectrum. 1-Chloropropane has **three** different types of carbon atom, in a methyl group (CH_3), a methylene group (CH_2) and a chloromethyl group (CH_2Cl); hence it will give three signals in its ^{13}C-n.m.r. spectrum.

The possibility that two different products might arise was commented upon over a century ago by Markovnikov, who proposed the simple rule, now known as Markovnikov's rule, that when a reagent HX adds to an alkene with an unsymmetric double bond, the hydrogen atom becomes attached to the carbon atom that already bears the most hydrogen atoms. It is now realized that the situation is not quite as simple as originally thought, but the rule applies reasonably to most alkenes having no other substituent groups.

The results can be explained satisfactorily in modern terms. To form 2-chloropropane or 1-chloropropane the initial attack on propene must take place, respectively at the 1- or 2-carbon atoms:

$$CH_3—CH=CH_2 \xrightarrow{H—Cl} \begin{cases} CH_3\overset{+}{C}HCH_3 \xrightarrow{Cl^-} CH_3CHClCH_3 \\ CH_3CH_2\overset{+}{C}H_2 \xrightarrow{Cl^-} CH_3CH_2CH_2Cl \end{cases}$$

In the first case, the positive charge rests on a carbon atom to which two other carbon atoms are attached. Such an atom is known as a **secondary carbon atom**, and the carbenium ion is called a **secondary carbenium ion**. In the second case, the positive charge is on a carbon atom which is attached to only one other carbon atom; this is called a **primary carbon atom** and provides a **primary carbenium ion**. A carbon atom to which three other carbon atoms are attached is similarly known as a **tertiary carbon atom**.

It is found that, in general, tertiary carbenium ions are more stable than secondary carbenium ions, which are in turn more stable than primary carbenium ions. The reason for this is associated with the fact that alkyl groups tend to be slightly electron-donating, i.e. a bond to an alkyl group may be represented as

$$-\overset{|}{\underset{|}{C}} \leftarrow \mathbf{R}$$

This means that alkyl groups tend slightly to neutralize the positive charge on a carbenium ion to which they are attached. Since tertiary carbenium ions have three alkyl groups attached to them, the neutralization of charge is more effective than in secondary or primary carbons with only two or one alkyl groups attached. Partial neutralization of the charge tends to make the carbenium ion more stable, and this in turn means that less energy is required for their formation.

Thus in the reaction of propene with hydrogen chloride formation of $CH_3\overset{+}{C}HCH_3$ requires less energy than formation of $CH_3CH\overset{+}{C}H_2$. This may be illustrated by energy profiles for these reactions:

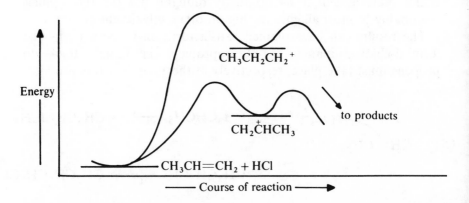

In consequence formation of the secondary carbenium ion and, hence, finally, of 2-chloropropane proceeds the more readily and is the predominant reaction. (Note that it is only the predominant reaction, not the only reaction; some few molecules may have enough energy to produce the primary carbenium ion. It is a common feature of organic chemistry that competing reactions may be possible, and although one reaction often predominates, it is less common for the other competing reactions to be completely absent. This is one reason why the percentage yields of products are commonly recorded in reports of organic reactions.)

Reactions of Alkenes with Hydrogen Bromide in the Presence of Light

When hydrogen bromide reacts with propene in the dark the major product is 2-bromopropane. If, however, reaction takes place in the presence of strong light, the main product is 1-bromopropane.

This change suggests that different chemical mechanisms may be involved in the two cases.

Light is known to promote chemical reactions by breaking the bonds in molecules. In doing this, the bonds are broken symmetrically with one electron going to each of the atoms which the bond has linked:

$$X \overset{\frown\frown}{\_} Y \xrightarrow{h\nu} X\cdot + \cdot Y$$

The products have unpaired electrons and are called **free radicals** or just **radicals**. Note three types of symbolism in the above reaction:

1. $h\nu$ is often used as a symbol for light.
2. The dot is used to represent an unpaired electron.
3. The 'fish hook', $\frown$, is used to indicate the movement of single electrons. Their use is analogous to that of curved arrows to denote the movement of pairs of electrons and the same conventions apply in their precise use, i.e. the tail of the hook indicates the source of the electron and the head the site to which it goes.

Cleavage of bonds symmetrically to give two radicals is known as **homolytic** cleavage of a bond. Cleavage in which both electrons go to one of the atoms which the bond linked, providing a negatively charged atom, while the other atom is left with a positive charge, viz.

$$X \overset{\frown}{\_} Y \rightarrow \overset{+}{X} + \bar{Y}$$

is known as **heterolytic** cleavage of a bond.

That radicals are involved in the reactions of alkenes with hydrogen bromide in the presence of light is also indicated by the fact that compounds known as 'radical initiators', which are known in general to promote reactions involving radicals, will induce the same type of reaction.

Reaction is likely to be started by homolytic cleavage of the weakest bond. In the system alkene + HBr, the weakest bond is H—Br. This will be broken to give bromine and hydrogen atoms:

$$Br \overset{\frown\frown}{\_} H \xrightarrow{h\nu} Br\cdot + H\cdot$$

Atoms or radicals with unpaired electrons are most usually highly reactive.

If a radical initiator is used to start the reaction, it commonly does this as follows; radical initiators are very often molecules which break very easily to provide radicals:

$$\widehat{X}\!-\!\widehat{X} \longrightarrow 2X\cdot$$

Radical $X\cdot \,\, \widehat{H}\!-\!\widehat{Br} \longrightarrow XH + Br\cdot$
initiator

$\widehat{Br}\cdot + \widehat{CH_2}\!\!=\!\!\widehat{CHCH_3} \longrightarrow BrCH_2\!-\!\dot{C}HCH_3$

$BrCH_2\!-\!\dot{C}HCH_3 + \widehat{H\!-\!Br} \longrightarrow BrCH_2CH_2CH_3 + Br\cdot$

Thus once this reaction is started, either by light or by an initiator, it is self-propagating, in other words it is a **chain reaction**, for in the final step another bromine atom is generated which can attack another propene molecule, and so on.

Reactions of radicals are usually very fast and very indiscriminate; they attack the nearest thing to hand. Where they attack is usually the most accessible site. In the case of propene this means attack on the carbon atom at the end of the molecule rather than on the carbon atom at the other end of the double bond. Hence 1-bromopropane is the principal (though not necessarily the only) product. Note that this is the opposite product from that predicted by Markovnikov's rule which does not apply to these radical addition reactions.

Hydrogen chloride and hydrogen iodide do not react with alkenes by radical mechanisms.

Chlorine and bromine may react with alkenes by a radical reaction in the presence of light and in the gas phase (or less commonly in a non-polar solvent). In polar solvents the ionic mechanism operates.

Reduction of Alkenes

Alkenes are reduced to alkanes by addition of hydrogen to the double bond. Alkenes do not react with hydrogen except in the presence of a suitable catalyst. This catalyst is usually a metal, for example platinum or palladium. Both hydrogen and alkene are absorbed on the surface of the catalyst which serves both to bring the molecules together to react and to lower the energy required for reaction to take place. An energy profile could be as follows:

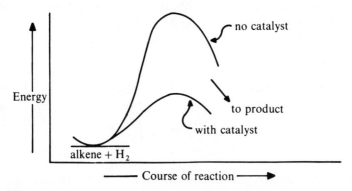

— Course of reaction —▶

Catalytic reduction, also called **catalytic hydrogenation,** of alkenes is of considerable industrial importance. Alkenes are formed in the cracking of oil (see Chapter 2) and are among the most important raw materials for use in the chemical industry. One such use can be shown, outrageously oversimplified, as

$$\text{oil} \xrightarrow{\text{cracking}} \text{alkenes} \xrightarrow[\text{catalyst}]{H_2} \text{alkanes, petrol}$$

Similarly margarine and cooking fats are made by catalytically hydrogenating vegetable oils, which contain alkene double bonds (see Chapter 17).

Chemical Tests for Alkenes

Common simple visual tests for the presence of alkenes are to add to separate portions of the sample under test (*a*) bromine and (*b*) an aqueous solution of potassium permanganate. Alkenes decolorize both of these reagents; the bromine adds to the double bond(s) as described above, the permanganate oxidizes the double bond. In each case the colour of the test reagent (bromine–brown, potassium permanganate–purple) disappears.

Oxidation of Alkenes

If an alkene is treated with a cold dilute solution of potassium permanganate in aqueous alkali, the permanganate first adds to the double bond to give a transient product which reacts with water to form a compound with two HO (hydroxy) groups on the carbon atoms which were initially linked by the double bond. Such 1, 2-dihydroxy compounds are sometimes called **glycols** (see Chapter 9):

$$RCH{=}CHR' + MnO_4^- \longrightarrow \left[\begin{array}{c} RCH{-}\!\!-\!\!CHR' \\ | \qquad\quad | \\ O \qquad\ O \\ \diagdown \quad \diagup \\ MnO_2 \end{array} \right]^- \xrightarrow{H_2O} \begin{array}{c} RCH{-}CHR' \\ | \qquad | \\ OH \quad OH \end{array}$$

If acid conditions are used the glycols are further oxidized, breaking a carbon–carbon bond, to give compounds called **carboxylic acids** (see Chapter 15):

$$\begin{array}{c} RCH{-}CHR' \\ | \qquad | \\ OH \quad OH \end{array} \xrightarrow[\text{H}_2\text{O, acid}]{\text{KMnO}_4} RCOOH + R'COOH$$

A different oxidation reaction takes place if hydrogen peroxide is used as the oxidizing agent:

$$RCH{=}CHR' \xrightarrow{H_2O_2} \begin{array}{c} RCH{-}\!\!-\!\!CHR' \\ \diagdown \quad \diagup \\ O \end{array}$$

These compounds, having three-membered rings made up of two carbon atoms and one oxygen atom are called **epoxides** or **oxirans** (see Chapter 10).

The epoxide obtained from ethylene is commonly called **ethylene oxide**, and is of enormous industrial importance (see Chapter 10). It is made industrially by treating ethylene with oxygen in the presence of a catalyst, commonly silver:

$$\begin{array}{c} CH_2 \\ \| \\ CH_2 \end{array} + O_2 \xrightarrow{\ Ag\ } \begin{array}{c} CH_2 \\ | \quad \diagdown \\ CH_2 \quad O \\ \diagup \end{array}$$

The **hardening of fats and of paints** involves formation of epoxides, by oxidation of double bonds present in the fats or paints by atmospheric oxygen.

Polymerization of Alkenes

Since alkenes readily undergo addition reactions with a variety of other kinds of molecules, it prompts the question as to whether alkene molecules might add to one another.

Under the appropriate conditions they do. In these addition reactions, schematically, if not mechanistically, one may compare them with addition reactions of other HX species. For example, with ethylene, $2C_2H_4 \rightarrow C_4H_8$ and this could be further expressed as $C_2H_3H + CH_2 = CH_2 \rightarrow C_2H_3CH_2 - CH_3$. Using structural formulae this becomes

$$CH_2{=}CH_2 + CH_2{=}CH_2 \longrightarrow CH_2{=}CH{-}CH_2{-}CH_3 \quad \text{But-1-ene}$$

Such a reaction, in which two identical molecules unite to form one molecule having exactly double the molecular formula is called **dimerization**, and the product is a **dimer** of the starting compound, which may be called the **monomer**.

The resultant dimer in this case still has a double bond and can hence add a third ethylene molecule, and so on, and a product made up of many molecules of the starting compound results, for example

$$C_2H_4 \xrightarrow{\;C_2H_4\;} C_4H_8 \xrightarrow{\;C_2H_4\;} C_6H_{12} \xrightarrow{\;nC_2H_4\;} (C_2H_4)_n$$

| Ethylene | But-1-ene | (Trimer) | Polymer |
| (monomer) | (dimer) | | |

Such a product is called a **polymer** (poly- signifies 'many') and the process is known as **polymerization**.

Polymerization is an extremely important industrial process, since many plastics which are important materials of commerce are polymers. Apart from industrial uses, many domestic implements are made from polyalkene plastics, e.g. buckets are commonly made from polyethylene or polypropene (sometimes called polypropylene). Many of these polymers can be moulded to give a wide variety of articles. Sometimes mixtures of different alkenes **copolymerize** to provide polymers with particular useful properties. Indeed by using different monomers and by using different conditions for the polymerization process, polymers with many different properties may be obtained and polymers may be tailor-made for different uses.

Polymerization does not take place readily unless it is catalysed, either by an acid or by radicals. The general equations for such reactions are as follows (A^+ = acid catalyst, $X^.$ = radical catalyst):

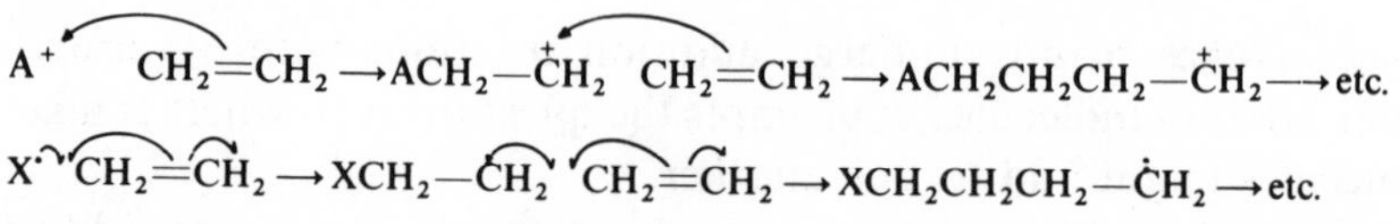

$A^+ \quad CH_2{=}CH_2 \rightarrow ACH_2{-}\overset{+}{C}H_2 \quad CH_2{=}CH_2 \rightarrow ACH_2CH_2CH_2{-}\overset{+}{C}H_2 \rightarrow$ etc.

$X^. CH_2{=}CH_2 \rightarrow XCH_2{-}\overset{.}{C}H_2 \; ^.CH_2{=}CH_2 \rightarrow XCH_2CH_2CH_2{-}\overset{.}{C}H_2 \rightarrow$ etc.

Polymerization ceases when the carbenium ion or radical is trapped by some other molecule present before it can add to another molecule of alkene. Control of the extent to which polymerization proceeds, i.e. to how many monomer units join together, is extremely important in producing polymers having the desired properties for the job for which they are required.

Two of the commonest monomers in use industrially are ethylene, which provides polyethylene (sometimes described as polythene), and propene, which provides polypropene (sometimes described as poly-propylene). Both of these alkenes, which are gases, are formed in the refining of oil. Two other common monomers which are alkenes are vinyl chloride ($CH_2{=}CHCl$) (or chloroethylene; the $CH_2{=}CH-$ group of atoms is called the vinyl group), which provides poly(vinyl chloride) (PVC), widely used for making piping, upholstery, wire insulation, etc., and tetrafluoroethylene ($CF_2{=}CF_2$) which is converted into poly(tetrafluoroethylene), Teflon, used for vessels, etc., which need to be highly resistant to chemicals and to heat (as Teflon is). Teflon is met domestically as a non-stick coating on frying pans and other kitchenware. Many other alkenes are also used industrially in the production of polymers.

Uses, Importance and Source of Alkenes

Ethylene is important not only in the production of plastics but also as the starting material from which a wide variety of organic compounds is made. It is one of the world's most important raw materials. Other alkenes are similarly important for the production both of plastics and organic compounds of many kinds. All are derived ultimately from the treatment of crude oil in oil refineries. Another important, if smaller, use of ethylene is in the storage of various fruits such as apples, bananas and citrous fruit; it makes their storage and ripening possible and enables them to be kept and transported with greater ease.

Questions

1. What are the major products to be expected from the reactions of propene with (a) hydrogen chloride, (b) bromine in aqueous solution, (c) bromine in solution in ethanol and (d) bromine in an aqueous solution of sodium chloride? Provide mechanisms for the reactions.
2. Write mechanisms for the addition reactions of propene with hydrogen bromide (a) in solution and in the absence of light and (b) in the gas phase and in light.
3. Complete the following reaction schemes:

$$CH_2{=}CH_2 + KMnO_4 \longrightarrow ?$$

$$CH_3CH{=}CH_2 + conc.\ H_2SO_4 \longrightarrow ? \xrightarrow{H_2O} ?$$

$$CH_3CH{=}CH_2 + Br_2 \xrightarrow{hv} ?$$

4. Draw energy profiles for the formation of primary and secondary carbenium ions from propene. Which is the preferred reaction?

6 REACTIONS OF ALKANES

Alkanes have no 'extra' electrons that are not essential for holding the molecules together. Hence they will not undergo the addition reactions typical of alkenes. For this reason they are described as **saturated compounds.** Furthermore, the C—C and C—H bonds which are present in alkanes are strong bonds and not easily broken. Additionally they are not appreciably polarized, so there are no electron-rich or electron-poor centres in the molecules that might be attacked, respectively, by electrophiles or nucleophiles such as H^+ or HO^- (or H_2O).

In consequence of these factors, alkanes are rather unreactive compounds; hence common alkanes such as natural gas, petrol, paraffin and paraffin wax may be kept for very long periods without undergoing change. The old chemical name for alkanes was **paraffins**, derived from the Latin words *parum*, little, and *affinis*, affinity, because they showed so little tendency to react with other molecules.

Although generally unreactive, some reactions which they undergo when suitably initiated involve a considerable release of energy. Examples are their reactions with oxygen, chlorine or fluorine. However, all these reactions need initiation, in that they need an initial input of energy to start them. Thus, although petrol may be kept indefinitely, application of energy, in the form of a lighted match, results in a violent reaction with the oxygen of air, with an evident large release of energy. The final products are carbon dioxide and water. The reaction between alkanes and chlorine is initiated by shining light onto a mixture of the two.

The initial input of energy, either heat or light, to start these reactions is required in order to provide sufficient energy to break some bonds in the reacting molecules. Since the chemical bonds in the reacting molecules are not dipolar in nature, it is a reasonable assumption that, when these bonds are broken, homolytic cleavage will take place, providing radicals.

Chlorination of Alkanes

Combustion (oxidation) of alkanes is a very complicated process, so let us rather consider first their reactions with chlorine. The simplest example is

$$CH_4 + Cl_2 \longrightarrow CH_3Cl + HCl$$
$$\text{Chloromethane}$$

It is most likely that the bonds first broken, under the influence of light, will be the weakest bonds. The Cl—Cl bond is a weaker bond than a C—H bond, so the initial reaction is likely to be

$$Cl—Cl \xrightarrow{h\nu} Cl\cdot + \cdot Cl$$

The resultant chlorine atoms are extremely reactive since they each have an unpaired electron, and will at once attack any adjacent molecule. Possible reactions are

(a) $Cl\cdot + Cl\cdot \longrightarrow Cl_2$

(b) $Cl\cdot + Cl—Cl \longrightarrow Cl—Cl + Cl\cdot$

(c) $Cl\cdot + CH_4 \longrightarrow HCl + \cdot CH_3$

Reaction (a) is statistically unlikely because at any instant there will be very few chlorine atoms present and available for reaction. (Furthermore, the chlorine molecule, Cl_2, which would be formed, would have a very high energy, which is produced in the exothermic reaction $2\,Cl\cdot \longrightarrow Cl_2 + $ energy, and because of this would very easily break again into chlorine atoms.) Reaction (b) only results in the exchange of one chlorine atom for another.

However, reaction (c) results in the formation of two new species, a methyl radical and hydrogen chloride, which is one of the final products of the overall reaction.

We must now consider further reactions of the methyl radical and its conversion into chloromethane. A straightforward reaction with a chlorine atom, as in

$$Cl\cdot + \cdot CH_3 \longrightarrow ClCH_3$$

is statistically unlikely, again because the numbers of such species present at any one time is exceedingly small (because of their great reactivity) and there is consequently little likelihood of their meeting.

Other possible and statistically much more likely reactions of a methyl radical are

$$\cdot CH_3 + CH_4 \longrightarrow CH_4 + \cdot CH_3$$
$$\cdot CH_3 + Cl_2 \longrightarrow CH_3Cl + \cdot Cl$$

The latter reaction provides the other final product of the overall reaction and at the same time produces a new chlorine atom, which can set the whole reaction process off again. Overall we have

$$Cl—Cl \xrightarrow{hv} Cl\cdot + \cdot Cl \qquad \text{Initiation}$$

$$\left. \begin{array}{l} \cdot Cl + CH_4 \longrightarrow HCl + \cdot CH_3 \\[2mm] \cdot CH_3 + Cl_2 \longrightarrow CH_3Cl + \cdot Cl \end{array} \right\} \text{Propagation}$$

Once this process is started it should continue indefinitely since, as well as the products, a further chlorine atom is produced; a **chain reaction** is set in motion. These steps are described as **initiation** and **propagation** steps, as indicated.

In theory only *one* initiation, by splitting one chlorine molecule, is needed to cause complete reaction of all the molecules present in such a chain reaction, but in fact there is always wastage from **recombination reactions** which cause **termination** steps, as follows:

$$\left. \begin{array}{l} Cl\cdot + \cdot Cl \quad \longrightarrow Cl_2 \\[2mm] \cdot CH_3 + \cdot CH_3 \longrightarrow CH_3—CH_3 \\[2mm] Cl\cdot + \cdot CH_3 \longrightarrow CH_3Cl \\[2mm] \cdot CH_3 \quad \text{or} \quad \cdot Cl + \text{wall of flask, etc.} \end{array} \right\} \text{Termination}$$

Although not widespread compared to the propagation reactions, they are sufficient to prevent only one initiation being adequate. The last of these termination reactions, involving extraneous molecules, is probably the most important.

If the energetics of the reaction are considered the initiation reaction requires input of energy, i.e. it is endothermic, but both the propagation reactions are exothermic, and the whole process is exothermic overall. In fact it is sufficiently exothermic that the reaction can be explosive. A mixture of methane and chlorine remains as such in the dark, but

if a light is shone on the mixture, it very often explodes. The energy
profile for the reaction is:

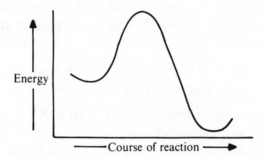

Hence, too, the reaction sustains itself once it is started. (The
endothermic or exothermic nature of a reaction may be deduced from
a consideration of the bond strengths of the reactants and the products,
but will not be discussed further here.)

The reaction of chlorine with an alkane may be catalysed by a
so-called initiator. The latter will be a compound with a relatively
weak chemical bond in its structure, which can be broken homolytically under fairly gentle conditions. Let us consider some such molecule
X—X [an example is $(CH_3)_3C$—O—O—$C(CH_3)_3$, whose O—O
bond is relatively weak] in which the bond X—X is readily broken
homolytically. This provides radicals X· which can react with chlorine
to initiate the usual chain reactions:

$$X—X \longrightarrow 2X·$$
$$X· + Cl_2 \longrightarrow ClX + ·Cl$$
$$\text{etc.}$$

Chlorination of an alkane does not necessarily stop when only one
hydrogen atom has been replaced by chlorine, and the extent of
substitution depends on the amount of chlorine available. Thus in the
chlorination of methane all the following products may be formed and
usually a mixture of them results:

$$CH_4 \xrightarrow{Cl_2} CH_3Cl \xrightarrow{Cl_2} CH_2Cl_2 \xrightarrow{Cl_2} CHCl_3 \xrightarrow{Cl_2} CCl_4$$

Chloro-methane (or methyl chloride)	Dichloro-methane (or methyl-ene chloride)	Chloroform (or trichloro-methane)	Carbon tetrachloride (or tetrachloro-methane)

All these products have unofficial names as well as the systematic ones, and in the cases of trichloro- and tetrachloro-methane, the commonly used names are chloroform and carbon tetrachloride, and these names must be known and recognized. For chloro- and dichloro-methane the alternative names are also common, but not universally so, as for the other two.

All alkanes react similarly to methane with chlorine but note that more than one isomer may (and probably will) be formed. For example, monochlorination of propane will provide 1-chloropropane and 2-chloropropane:

$$CH_3CH_2CH_3 + Cl_2 \begin{cases} CH_3CH_2CH_2Cl \\ CH_3CHClCH_3 \end{cases}$$

Reactions of other Halogens with Alkanes

Bromine and fluorine also react with alkanes but iodine does not. Fluorine reacts even more vigorously than does chlorine; the reaction needs no initiation, and when fluorine and a gaseous alkane are mixed, an explosion occurs spontaneously. Bromine reacts more sluggishly.

The differences in reactivity are explicable in terms of the energetics, or energy changes, involved in the reactions.

In the case of bromine the situation is as follows:

$$Br{-}Br \longrightarrow Br\cdot + \cdot Br \qquad \text{Endothermic}$$
$$\cdot Br + CH_4 \longrightarrow HBr + \cdot CH_3 \qquad \text{Endothermic}$$
$$\cdot CH_3 + Br_2 \longrightarrow CH_3Br + \cdot Br \qquad \text{Exothermic}$$
$$\text{Total overall reaction—Exothermic}$$

Thus in this case both of the first two steps are endothermic and more energy is required to initiate the reaction. The overall reaction is exothermic and so is self-propagating, but is less so than in the case of the reaction with chlorine.

In the case of fluorination both propagation steps are highly exothermic (they lead to the formation of very strong H—F and C—F bonds), and the reaction as a whole is strongly exothermic.

Oxidation of Alkanes

The reaction of alkanes with oxygen is much more complicated than the reaction with chlorine. It too is highly exothermic. This is why the combustion of alkanes, i.e. of gas, petrol and oil, is so important in everyday life and why they are such valuable fuels. The idealized equation for the complete combustion of methane is

$$CH_4 + 2O_2 \longrightarrow CO_2 + H_2O + 882\,kJ/mol\ (= 55\,kJ/g)$$

(This figure of 882 kJ/mol may be compared with the energy normally available from the environment, which amounts to ~ 60–$80\,kJ$.)

In fact combustion often does not proceed to completion, and by incomplete combustion carbon and/or carbon monoxide are obtained. Both of these products are of considerable industrial importance.

Carbon may be obtained in finely divided form known as carbon black. This has use in inks, as a pigment, in tyres, as a reinforcing agent to increase their strength, and in the making of carbon fibres.

Carbon monoxide is one of the most important of all starting materials in industry. To take only a few of many important applications, by reaction with hydrogen it is converted into methanol (CH_3OH) (see Chapter 9) and thence into formaldehyde (HCHO) (see Chapter 14), both extremely useful chemicals; by reaction with chlorine into phosgene (carbonyl chloride, $COCl_2$) and thence into polyurethane foams; and by addition to alkenes to provide a variety of aldehydes (see Chapter 14).

A rough equation for the conversion of methane into carbon monoxide is

$$2CH_4 + O_2 \longrightarrow 2CO + 4H_2$$

From this it may be seen that a by-product is hydrogen and this is yet another important feedstock for the chemical industry.

Biological Oxidation of Alkanes

Certain bacteria and yeasts contain **enzymes**, which are very specialized catalysts occurring naturally in living matter, and these particular enzymes can oxidize alkanes at room temperature. These bacteria and yeasts can be fed on mineral oil and air. Alkanes are thus converted into very complex organic molecules, including fats, sugars and, if a source of nitrogen is also provided, proteins. By this means a source of synthetic foodstuffs from mineral oil is available. The chemistry involved is much too complex to be fully understood.

Question

1. Does ethane react with (a) chlorine and (b) potassium chloride? If a reaction does take place, show the mechanism. If reaction does not take place, explain why this is so.

7 ALKYL HALIDES. SOME EFFECTS OF SOLVENTS

Alkyl halides are derived from alkanes by replacing a hydrogen atom with a halogen atom. If Hal = halogen their general formula is $C_nH_{2n+1}Hal$ or $RHal$.

In preceding chapters it has been shown that alkyl halides may be formed by the reactions of alkanes with chlorine or bromine, or by the addition of hydrogen halides to alkenes. They are commonly made in the laboratory from alcohols; this will be considered in Chapter 9.

Nomenclature

The systematic nomenclature is as discussed in previous chapters, and the halides are named as halogeno-alkanes. The halogen atoms are designated by the prefixes fluoro, chloro, bromo and iodo. Examples are

$CH_3CHClCH_2CH_3$ 2-Chlorobutane

⬡—Br   Bromocyclopentane

$CH_2F-CH_2-CH-CHI-CH_3$ 1-Fluoro-4-iodo-3-methylpentane
                  |
                CH_3

Note that substituents are listed in alphabetical order, not numerical order; this is done to assist in indexing names of compounds.

If the halogen atom is attached to the carbon atom at the end of a chain of carbon atoms, the so-called terminal carbon atom, the compound may also be named in another way, for example CH_3CH_2Cl, chloroethane, may be called ethyl chloride. In other words the compound is named as an alkyl halide rather than as a halogeno-alkane. In this form of nomenclature we have

CH_3Cl	Methyl chloride
CH_3CH_2Br	Ethyl bromide
$CH_3CH_2CH_2I$	Propyl iodide

However, for any compound in which a halogen atom is attached to a non-terminal carbon atom the other form of nomenclature should be used, that is CH_3CHICH_3 should be called 2-iodopropane. (This compound is, however, sometimes called isopropyl iodide, the CH_3CHCH_3 group being known as the isopropyl group.)

Note that propyl iodide ($= 1$-iodopropane) and 2-iodopropane are isomers. Similarly there are four possible alkyl chlorides with molecular formula C_4H_9Cl, namely

(a) $CH_3CH_2CH_2CH_2Cl$ 1-Chlorobutane or butyl chloride
(b) $CH_3CH_2CHClCH_3$ 2-Chlorobutane
(c) CH_3—CH—CH_2Cl 1-Chloro-2-methylpropane
　　　　　| (or isobutyl chloride)
　　　CH_3
(d) $(CH_3)_3CCl$ 1-Chloro-1,1-dimethylethane
　　　　　　　　　　　　　　　　(or tertiary butyl chloride
　　　　　　　　　　　　　　　　or t-butyl chloride)

Isomers (c) and (d) both have their chlorine atoms attached to terminal carbon atoms. The groups $(CH_3)_2CHCH_2$— and $(CH_3)_3C$— have the permitted trivial names isobutyl and t-butyl (t = tertiary); hence (c) and (d) may alternatively be described as isobutyl chloride and tertiary butyl chloride.

The name tertiary butyl derives from the fact that the carbon atom to which the substituent chlorine atom is attached is a tertiary carbon atom, as defined in Chapter 5. Of the above isomers, (d) is a tertiary halide, (a) and (c) are primary halides, and (b) is a secondary halide since the halogen atoms are attached, respectively, to tertiary (d), primary (a and c), and secondary (b) carbon atoms.

General Characteristics. Nature of Carbon–Halogen Bonds

Most alkyl halides having small alkyl groups are liquids which are either not very soluble or insoluble in water.

The molecules are weakly polar because the electrons in the carbon–halogen bond are not equally shared between the carbon and halogen atoms but tend to be associated more with the halogen atom, as discussed at the end of Chapter 2. This may be represented by the symbol $C \rightarrow Cl$. It may also be expressed as $\overset{\delta +}{C} - \overset{\delta -}{Cl}$, indicating that the carbon atom is slightly deficient in electrons, or **electron-poor**, while the chlorine atom is slightly **electron-rich**. However the bonding is covalent and alkyl halides are not ionic compounds like metal halides such as sodium chloride.

Reactions of Alkyl Halides

Because carbon–halogen bonds have dipolar character, they tend to break heterolytically in reactions in solution.

Also, because the carbon atom bears a partial positive charge, reactions frequently involve attack by a nucleophile on the carbon atom. Because of their ready reactivity towards a large range of nucleophiles, alkyl halides are highly important reagents. Alkyl chlorides, bromides and iodides all react similarly, and in following examples either of these may be used.

One example of such a reaction is

$$Na^+ \quad HO^- + CH_3Cl \longrightarrow HOCH_3 + Cl^- \quad Na^+$$

In general,

$$Na^+ \quad HO^- + RHal \longrightarrow HOR + Hal^- \quad Na^+$$

The products ROH are known as **alcohols**; CH_3OH is methanol.

The sodium ion takes no part in the reaction, which involves only the hydroxide ion. It is therefore common practice, which will be followed in this book, only to write the ion which is actually involved in the reaction. Thus the above reaction becomes:

$$HO^- + CH_3Cl \longrightarrow HOCH_3 + Cl^-$$

It **must always** be remembered that the anion will not be present alone; there must always be a counterion present bearing a positive charge, even if it is not written out. It is useless to look for a bottle containing only hydroxide ions; the reagent is sodium hydroxide, or potassium hydroxide, etc.

Other anions react similarly; for example

$$RCl + HS^- \longrightarrow RSH + Cl^-$$
$$RBr + CN^- \longrightarrow RCN + Br^-$$

The products RSH are called **thiols**; the products RCN are called **alkyl cyanides** or **nitriles**. This reaction with cyanide ions is important because it creates a new carbon–carbon bond; for example

$$H_3C{-}Cl + CN^- \longrightarrow H_3C{-}CN + Cl^-$$

Methods of forming carbon–carbon bonds play a very important role in synthetic organic chemistry.

All the anions mentioned are derived from weak acids, HO^- from H_2O, HS^- from H_2S and CN^- from HCN. In general anions derived from weak acids react readily with alkyl halides in similar substitution reactions. Let us consider some other examples.

Organic **carboxylic acids**, like **acetic acid**, CH_3COOH, are weak acids having the general formula RCOOH (see Chapter 15). In solution only a small fraction of the molecules are dissociated into ions at any one time:

$$CH_3COOH \rightleftharpoons CH_3COO^- + H^+$$

However, salts of these acids, e.g. sodium acetate, are completely ionized, and the acetate ions react with alkyl halides:

$$CH_3COO^- + RBr \longrightarrow CH_3COOR + Br^-$$

The product, which is derived from a carboxylic acid by replacing the acidic hydrogen atom with an alkyl group, is called an **ester**; these compounds will be considered further in Chapter 17.

Alcohols, ROH, mentioned earlier in this chapter, are very weakly acidic, and can be regarded as not ionized. However, they react with sodium (and other metals) to give **alkoxides**:

$$2\,Na + 2\,ROH \longrightarrow 2\,RO^- \quad Na^+ + H_2$$

(cf. water + sodium → sodium hydroxide). Alkoxide ions resemble hydroxide ions and, like them, react with alkyl halides. For example,

$$CH_3O^- + CH_3Cl \longrightarrow CH_3OCH_3 + Cl^-$$
$$C_2H_5O^- + CH_3Br \longrightarrow C_2H_5OCH_3 + Br^-$$

or, in general,

$$RO^- + R'Br \longrightarrow ROR' + Br^-$$

These products, which will be considered later in Chapter 10, are known as **ethers**.

Anions derived from acids of moderate strength, e.g. halide ions, react much less readily in such reactions. Anions derived from very strong acids, e.g. sulphate or perchlorate from sulphuric acid or perchloric acid, do not react at all.

The reactivity of anions as nucleophiles and the strengths of the acids from which the anions are derived are related properties. Strong acids are strong because the anions show little tendency to react with positively charged protons to form undissociated acid molecules. Similarly these anions show little or no tendency to react with positively charged carbon atoms.

In the reactions so far described between anions and alkyl halides the reaction involves the transfer of a pair of electrons from the anion to the carbon atom of the alkyl halide:

$$[H-\ddot{O}:]^- \quad \overset{\delta+}{C}H_3 - \overset{\delta-}{C}l \longrightarrow H-O-CH_3 + Cl^-$$

Also in the reaction the chlorine atom leaves the carbon atom, taking with it the electrons that formed the carbon–chlorine bond, and becomes a chloride anion.

Neutral molecules having lone pairs of electrons not involved in chemical bonding might be expected to react similarly. An example of such a reaction involves water as the nucleophile:

$$
\begin{array}{c}
H \\
\diagdown \\
\ddot{O}: \quad CH_3 - Br \longrightarrow \\
\diagup \\
H
\end{array}
\qquad
\begin{array}{c}
H \\
\diagdown \\
\overset{+}{O} - CH_3 + Cl^- \\
\diagup \\
H \\
\diagdown \\
H^+ + HOCH_3
\end{array}
$$

Thus water also reacts with alkyl halides to form alcohols. Not surprisingly, reaction is much slower than the reaction when hydroxide ions are involved. Hydroxide ions bear a full negative charge and interact much more strongly with the slightly positively charged carbon atom of the alkyl halide than do neutral water molecules.

Ammonia, which also has a lone pair of electrons, reacts similarly to water with alkyl halides:

$$H_3N: \overset{\frown}{} CH_3 \overset{\frown}{} Br \longrightarrow H_3 \overset{+}{N} CH_3 \quad Br^-$$

It is, however, much more reactive than water. This parallels the fact that ammonia is more basic than water, i.e. it reacts more readily than water with protons.

The product above is methylammonium bromide, and is an example of an **alkylammonium salt**. As happens with ammonium salts, treatment with a base removes a proton, leaving an **amine** (see Chapter 8):

$$\overset{+}{N}H_4 \quad Cl^- \xrightarrow{B^-} BH + NH_3$$

$$CH_3 \overset{+}{N} H_3 \quad Br^- \xrightarrow[(B^- = base)]{B^-} BH + CH_3NH_2$$

All of the above reactions are examples of **substitution reactions**. Since the reactions involve attack of a nucleophile on the alkyl halide they are called **nucleophilic substitution reactions**.

Mechanism of Nucleophilic Substitution Reactions

Let us now consider how these nucleophilic substitution reactions take place. As an example we will take the reaction of hydroxide ions with methyl chloride:

$$HO^- + CH_3Cl \longrightarrow CH_3OH + Cl^-$$

One could envisage three ways in which this reaction might proceed, as follows:

1. Methyl chloride dissociates into methyl and chloride ions; the hydroxide ion then adds to the methyl ion:

$$CH_3Cl \longrightarrow \overset{+}{C}H_3 + Cl^-$$
$$\overset{+}{C}H_3 + HO^- \longrightarrow CH_3OH$$

2. The hydroxide ion adds to the methyl chloride to give a species $[HOCH_3Cl]^-$ which in turn loses a chloride ion:

$$HO^- + CH_3Cl \longrightarrow [HOCH_3Cl]^- \longrightarrow HOCH_3 + Cl^-$$

3. The hydroxide ion approaches a methyl chloride molecule, attracted by the partial positive charge on the carbon atom. As the hydroxide

ion gets nearer to the carbon atom it repels the chlorine atom, since the latter bears a partial negative charge, and eventually displaces it from the carbon atom:

$$HO^- + \overset{\delta+}{C}H_3 \overset{\delta-}{-}Cl \rightarrow \overset{\delta-}{H}O \cdots \overset{\delta+}{C}H_3 \cdots \overset{\delta-}{C}l \rightarrow HOCH_3 + Cl^-$$

Of these possible **mechanisms** for the reaction, mechanism 1 is unlikely because methyl chloride does not normally dissociate into ions and method 2 is unlikely because the carbon atom in the intermediate step would have five groups attached to it, whereas the normal covalency of carbon is four.

This leaves mechanism 3 as the only one not readily ruled out. If this mechanism is correct, the rate at which the reaction proceeds should be associated with the number of times hydroxide ions and methyl chloride molecules meet. The more opportunities there are for them to do so, the faster should be the reaction.

The chance of such meetings depends directly on the concentration of the reagents in solution; if the concentration of hydroxide ions is doubled so will be the rate of reaction. Likewise, doubling the concentration of methyl chloride will also double the rate of reaction.*

Therefore rate of reaction $\propto [HO^-]$ and rate of reaction $\propto [CH_3Cl]$. (The square brackets signify that we are referring to **concentrations** of the species within the brackets.)   Therefore,

$$\text{Rate of reaction} \propto [HO^-][CH_3Cl]$$

or

$$\text{Rate of reaction} = k[HO^-][CH_3Cl]$$

where k is a **rate constant** for the particular reaction involved.

Experiment shows this to be the case, providing strong evidence that mechanism 3 is indeed the mechanism involved.

Similar results are obtained for reactions of ethyl chloride, C_2H_5Cl, but **not** for all alkyl halides.

The rates of many reactions involving t-butyl chloride are dependent only upon the concentration of the alkyl halide and not on the

---

*A parallel for this argument can be drawn with your chances of having your pocket picked. Someone has to make contact with you to do this. There is more chance of someone making contact with you the more people you are among—the greater the concentration of people—i.e. in a crowd.

concentration of the nucleophile, that is

$$\text{Rate of reaction} = k[(CH_3)_3CCl]$$

Therefore the rate in these cases cannot be affected by the numbers of collisions between t-butyl chloride molecules and the nucleophile, but must depend upon some process involving only the halide. The only such process that can be considered likely is some ionization of the halide:

$$(CH_3)_3CCl \rightleftharpoons (CH_3)_3C^+ + Cl^-$$

If this ionization occurs only to a small extent the resultant carbenium ion will be very reactive and will at once either recombine with a chloride ion or, if an alternative nucleophile is present, with that nucleophile to give a new product. As the anion derived from a moderately strong acid, chloride is not a very reactive nucleophile, and the other nucleophile may well be more reactive.

Thus the suggested mechanism is

$$(CH_3)_3C\!-\!Cl \rightleftharpoons (CH_3)_3C^+ + Cl^-$$

$$(CH_3)_3C^+ + X^- \longrightarrow (CH_3)_3CX$$

$$(X^- = \text{nucleophile})$$

The second step will be much faster than the first, which involves the relatively sluggish ionization of the halide. When a number of reactions take place one after another, the overall rate must be dependent on the slowest step. This is called the **rate-determining step**.* In the example under consideration the rate-determining step is thus the dissociation of the t-butyl chloride into ions. The nucleophile is not involved in this, so the rate of reaction will be expressed by the equation

$$\text{Rate of reaction} = k[(CH_3)_3CCl]$$

Why does the reaction of t-butyl chloride go by this mechanism? Methyl chloride does not normally dissociate into ions but in the case of t-butyl chloride there is a slight amount of dissociation. It was mentioned in Chapter 5 that tertiary carbenium ions are more stable than primary carbenium ions. Since formation of carbenium and halide

---

*When you go, for example, to a football match, you may hurry to get to the ground and, once inside, hurry to the best vantage point. The slowest part is getting through the turnstiles. That is the rate-determining step of your progress.

ions from alkyl halides is an equilibrium

$$R - X \; \rightleftharpoons \; R^+ + X^-$$

$$(X = \text{halogen})$$

the more stable the resulting carbenium ion the more this equilibrium will favour its formation.

Primary carbenium ions have insufficient stability to enable them to be formed to any extent from alkyl halides, but the greater stability of tertiary carbenium ions results in some slight ionization taking place. Once formed, these carbenium ions are very reactive towards nucleophiles and, although ionization may only be slight, it will be sufficient to enable reaction with the nucleophile to take place. This in turn removes carbenium ions, and to restore the balance of the equilibrium more alkyl halide dissociates. The newly formed carbenium ions react with the nucleophile, and so the sequence continues until all the alkyl halide has been used up and the reaction is complete.

Secondary alkyl halides are intermediate in properties between primary and tertiary halides; for the most part they react by the same mechanism as methyl halides.

A reaction such as that of methyl halides with nucleophiles, which involves two interacting species (which may be neutral molecules and/or ions) in the rate-determining step, is known as a **bimolecular reaction**. A reaction such as that of t-butyl halides with nucleophiles, which involves only one species in its rate-determining step, is known as a **unimolecular reaction**.

These two types of mechanistic pathway for nucleophilic substitution reactions are described in shorthand form as S_N1- and S_N2-type reactions, referring, respectively, to the unimolecular and bimolecular reactions. S_N stands for substitution, nucleophilic, and 1 and 2 refer to **uni**- and **bi**-molecular respectively.

Mechanistically they can be summarized as follows (note again the precise use of the curved arrows):

S_N1: $R \overset{\frown}{-} Hal \rightleftharpoons R^+ + Hal^-$

                       $X^- + \overset{\frown}{R^+} \longrightarrow X - R$

S_N2: $X^- + \overset{\frown}{R} \overset{\frown}{-} Hal \longrightarrow X - R + Hal^-$

                 $(X^- = \text{nucleophile, Hal} = \text{halogen})$

These two mechanisms will have different energy profiles since S_N1 reactions have two separate steps whereas S_N2 reactions take place in one step. These may be represented as follows:

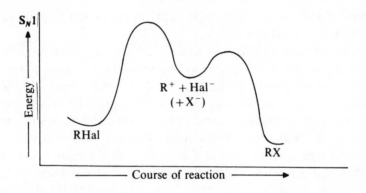

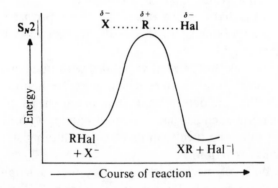

Reactions of alkyl halides all appear to be identical if they are thought of only in terms of the starting materials and products; the following equation in these terms is a general one:

$$RHal + X^- \longrightarrow RX + Hal^-$$

We have seen that these reactions are **not** all identical; some proceed by an S_N2 mechanism, others by an S_N1 mechanism. These reactions thus provide a salutary warning that reactions which may appear on paper to be similar may **not** in fact all go by the same mechanism. On a wider scale this example provides an excellent warning against making glib generalizations on the basis of insufficient information. Faced with the knowledge that both methyl and ethyl halides reacted by bimolecular mechanisms it could have been very easy to jump to the conclusion that all reactions of alkyl halides proceeded by the same mechanism. The temptation to extrapolation without evidence is always dangerous.

In fact, most reactions of alkyl halides may, in theory, go by S_N1 or S_N2 or both mechanisms, but, in practice, in very many cases one mechanism is so predominant that it alone appears to operate, and it is therefore justifiable to describe the result as an S_N1 or S_N2 reaction.

Effects of Solvent on Reactions

It is sometimes possible to influence the mechanism of a reaction by the nature of the solvent in which the reaction takes place.

In S_N1 reactions charged species, ions, are formed in the first, rate-determining step. Consequently any solvent that tends to stabilize ions will also tend to affect this step. Water can stabilize ions by solvation:

$$\text{H} \diagdown \ddot{\text{O}}\text{:} \cdots\cdots\cdot \text{R}^{+} \diagup \text{H}$$

In the case of carbenium ions, solvation may also lead to full bond formation:

$$\text{H} \diagdown \ddot{\text{O}}\text{:} \cdots\cdots \overset{|}{\underset{|}{\text{C}}}\overset{+}{-} \rightarrow \quad \text{H} \diagdown \overset{+}{\text{O}} - \overset{|}{\underset{|}{\text{C}}} - \rightarrow \quad \text{HO} - \overset{|}{\underset{|}{\text{C}}} -$$

A non-polar solvent, such as an alkane, cannot interact with an ion in this way.

Hence a polar solvent will help to stabilize a carbenium ion, compared to the effect of a non-polar solvent. Therefore use of a polar solvent will assist S_N1-type reactions. This can be illustrated by the energy diagram on the next page.

If this effect is big enough it may favour an S_N1 mechanism at the expense of a competing S_N2 mechanism.

Most reactions are carried out in solution. Roles of the solvent include the following:

(a) to serve as a medium for mixing the reagents,

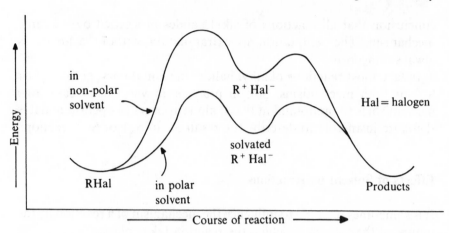

(b) to dissipate heat evolved from exothermic reactions,
(c) to assist in the control of temperature,
(d) to assist reaction (if chosen correctly) by providing suitable solvation.

It is self-evident that it is usually preferable to have a solvent in which both reactants dissolve. In the case of nucleophilic substitution reaction of alkyl halides the solubility of the reagents is frequently as follows:

alkyl halides—soluble in non-polar organic solvents, insoluble in water;
nucleophiles—insoluble in non-polar organic solvents, soluble in water.

The compromise situation is to use a polar organic solvent in which both reactants dissolve, very often an alcohol, ROH. Alcohols, like water, have lose pairs of electrons which might also react with the alkyl halide to give an unwanted product; for example

$$R \diagdown \overset{\cdot\cdot}{O} \diagup CH_3 \diagup Br \longrightarrow R \diagdown \overset{+}{O} - CH_3 \longrightarrow ROCH_3 + HBr$$
$$H \diagup \qquad\qquad H \diagup$$
$$\qquad\qquad\qquad\qquad\qquad Br^-$$

This is not, however, usually a problem, because there is a huge difference in reactivity between the anionic reagent and the neutral solvent, the latter being very much less reactive.

Sometimes a solvent can also be used more subtly. Consider the reaction of an alkyl chloride with sodium iodide:

$$Na^+ \quad I^- + RCl \rightarrow RI + Na^+ \quad Cl^-$$

We can also have a reaction of an alkyl iodide with sodium chloride:

$$Na^+ \quad Cl^- + RI \rightarrow RCl + Na^+ \quad I^-$$

Indeed, this can be regarded as an equilibrium:

$$NaI + RCl \rightleftharpoons RI + NaCl$$

The equilibrium usually favours the left-hand side in the above equilibrium. A carbon–chlorine bond is stronger than a carbon–iodine bond; associated with this an alkyl iodide is more reactive than the corresponding chloride. In consequence the equilibrium above lies on the side of the alkyl chloride.

If, however, a solvent is used in which sodium iodide is soluble but sodium chloride is not, the latter will precipitate out of solution and consequently the equilibrium is entirely displaced towards that side. Such a solvent is acetone, CH_3COCH_3. The situation is then as follows:

$$NaI + RCl \xrightleftharpoons{acetone} NaCl\downarrow + RI$$

Thus, by using the appropriate solvent the conversion of an alkyl chloride into an alkyl iodide can be achieved.

A carbon–fluorine bond is even stronger than a carbon–chlorine bond. Thus it could be expected that the following equilibrium would favour the alkyl fluoride:

$$RCl + F^- \rightleftharpoons RF + Cl^-$$

This is the case in very weakly polar solvents, but in alcohols this reaction, the conversion of an alkyl chloride into an alkyl fluoride, does not proceed. This is because in an alcohol the fluoride ion is very strongly solvated, so much so that it loses its nucleophilic reactivity.

It was mentioned above that anions derived formally from strong acids are very poor nucleophiles. Such anions may be made to react by using them in the form of their silver salts. The normal equilibrium is totally altered because silver halides are very insoluble and precipitate out of solution:

$$RCl + Na^+ \quad ONO_2^- \rightleftharpoons RONO_2 + Na^+Cl^-$$

$$RCl + Ag^+ \quad ONO_2^- \rightleftharpoons RONO_2 + AgCl\downarrow$$

By the use of methods exemplified in the foregoing examples, control over the course of reactions may be effected by the correct choice of solvent.

Elimination Reactions

A further complication may sometimes arise, in particular when strongly basic nucleophiles, such as HO^- or RO^-, are used. Nucleophilic substitution may be accompanied by, or even superseded by, a quite different reaction, for example

$$C_2H_5I + C_2H_5O^- \quad Na^+ \begin{array}{l} \nearrow C_2H_5OC_2H_5 \\ \searrow HOC_2H_5 + CH_2{=}CH_2 + Na^+ \quad I^- \end{array}$$

Hydrogen iodide has been extracted, or eliminated, from the ethyl iodide; such a reaction is described as an **elimination reaction.**

As with nucleophilic substitution, so elimination reactions may happen by two alternative mechanisms:

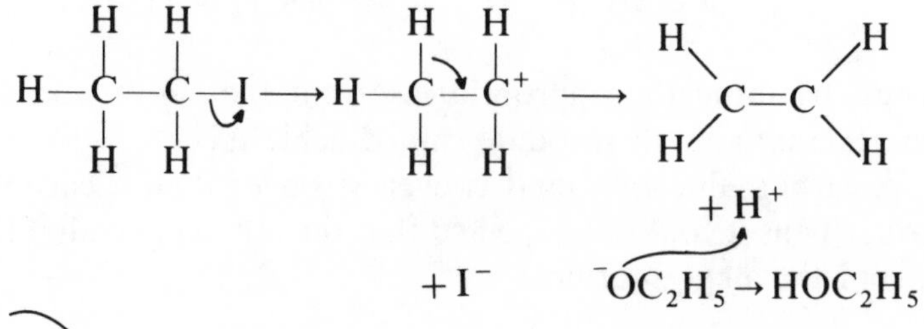

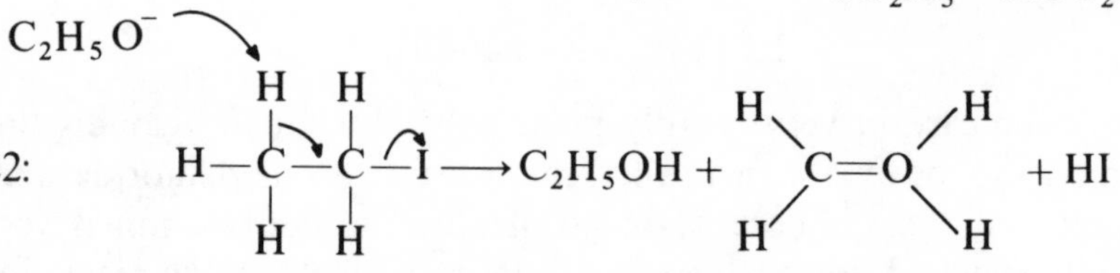

The shorthand names again reflect whether the reactions are unimolecular (E1) or bimolecular (E2) in the rate-determining step. E signifies elimination.

In the example given, it is actually the E2 mechanism that operates, and this is probably the more common mechanism.

The E1 mechanism, like the S_N1 mechanism, involves the formation of a carbenium ion, and is thus more likely to occur for a tertiary alkyl halide. However, even in the case of tertiary halides, the E2 mechanism is common, especially so the more strongly basic is the nucleophile.

Note that in E1 or E2 reactions the eliminated hydrogen atom comes from the carbon atom next to the one to which the halogen atom is attached. This is often described as the β-carbon atom, letters of the Greek alphabet being used to signify the positions of the carbon atoms, starting from that bearing the substituent group, which is designated α; for example

$$X-\overset{\alpha}{C}H_2-\overset{\beta}{C}H_2-\overset{\gamma}{C}H_2-\overset{\delta}{C}H_3$$

If there are no hydrogen atoms attached to the β-carbon atom, elimination of HX cannot take place. As an example

$$CH_3-\underset{\underset{\displaystyle CH_3}{|}}{\overset{\overset{\displaystyle CH_3}{|}}{C}}-CH_2Cl \xrightarrow{X^-} CH_3-\underset{\underset{\displaystyle CH_3}{|}}{\overset{\overset{\displaystyle CH_3}{|}}{C}}-CH_2X \text{ and } \textbf{no} \text{ elimination}$$

$$(X^- = \text{nucleophile})$$

Elimination of hydrogen halide from an alkyl halide often provides a useful method for the preparation of some alkenes, but the possibility of competing substitution reactions has always to be borne in mind. Especially, substitution predominates in the case of primary halides with unbranched chains of carbon atoms.

Whenever an alkyl halide is treated with a nucleophile there is the possibility of competition between two different reactions, substitution and elimination, and sometimes between more than two.

This is a very common feature in organic chemistry. Unlike, for example, simple acid–base titrations, organic reactions do not usually go completely to give the desired product, but the intended products are more often than not accompanied by alternative products, described as **by-products**.

This is a reason why the **percentage yields** of organic reactions, i.e. the percentage actually obtained of the theoretically possible yield if all the starting products were transformed into the desired products, are quoted. This provides a measure of the efficiency of the process as a means of preparing the compound concerned. It is also a reason why there is less insistence on balanced equations in organic chemistry, since they may give false expectations.

It is also because reactions of organic compounds may vary with the conditions or precise reagent used that it is necessary to follow precisely the recipes for reactions when they appear in print. Often the conditions to give the highest possible yield of the desired product have been carefully worked out, and, if these conditions are not adhered to, by-products may be formed instead.

If the reactions of ethyl iodide with a nucleophile are considered, namely

$$X^- + C_2H_5I \nearrow \begin{matrix} C_2H_5X + I^- \\ \\ C_2H_4 + HI \end{matrix}$$

$$(X^- = \text{nucleophile})$$

then the following observations apply. With most nucleophiles the substitution reaction is favoured. The more basic the nucleophile is, the more likely it is to remove H^+, and hence the more likelihood there is of some elimination reaction taking place. Of the nucleophiles previously mentioned in this chapter, the basicity is in the order $C_2H_5O^- > HO^- > RCOO^- > NH_3 > H_2O$, so that the amount of elimination should decrease down this series. With water no elimination takes place.

For all alkyl halides, in order to carry out elimination reactions sodium ethoxide would be used rather than sodium hydroxide. Better, an even stronger base, such as sodium amide, $Na^+NH_2^-$, would be used. Hence some textbooks used to say that the reaction of alkyl halides with hydroxide ions in aqueous solution gave alcohols, but in solution in ethanol, where an equilibrium

$$HO^- + C_2H_5OH \rightleftharpoons C_2H_5O^- + H_2O$$

provides some ethoxide ions, the reaction gave alkenes.

Tertiary halides tend more to elimination reactions than do secondary or primary alkyl halides unless the latter have branched carbon chains, especially when the branching occurs at the β-carbon atom.

These factors influence the way in which a reaction may be carried out. For example, to make the ether $CH_3OCH(CH_3)_2$, on paper one might use either of the following reactions:

(a)

$$\underset{\underset{\displaystyle CH_3}{|}}{\overset{\overset{\displaystyle CH_3}{|}}{H-C-O^-}} \quad Na^+ + CH_3Cl$$

$$CH_3-O-\underset{\underset{\displaystyle CH_3}{|}}{\overset{\overset{\displaystyle CH_3}{|}}{C}}-H$$

(b) $CH_3O^- \quad Na^+ + Cl-\underset{\underset{\displaystyle CH_3}{|}}{\overset{\overset{\displaystyle CH_3}{|}}{C}}{-}{-}H$

It is obviously better to use the first of these reactions. In reaction (b) elimination is a possible, and probable, alternative reaction; it is not possible in reaction (a), where the methyl chloride has no β-hydrogen atoms (or β-carbon atom).

Alkyl halides take part in these reactions, substitution and/or elimination reactions, very readily because the halide ions that are formed are stable, and in consequence are formed readily. They are described as good **leaving groups**. Hence alkyl halides are very valuable reagents in organic chemistry.

Uses of Alkyl Halides

Because of this reactivity alkyl halides are used not only in the laboratory but also industrially. Chlorine is very much cheaper than bromine or iodine and so in industry alkyl chlorides are most frequently used. Alkyl bromides and alkyl iodides are more reactive than alkyl chlorides, so far laboratory scale work, where smaller quantities are used and hence costs may be less vital, bromides and iodides are more commonly used.

Polyhalogeno Compounds

Compounds having two halogen atoms attached to one carbon atom are less reactive towards nucleophiles than are alkyl halides. Consequently they may be useful solvents. Thus dichloromethane reacts much less readily with hydroxide ions than does methyl chloride. The product is formaldehyde, CH_2O. The initial reaction to give chloro-

A First Course in Organic Chemistry

hydroxymethane is followed by an elimination reaction, providing formaldehyde:

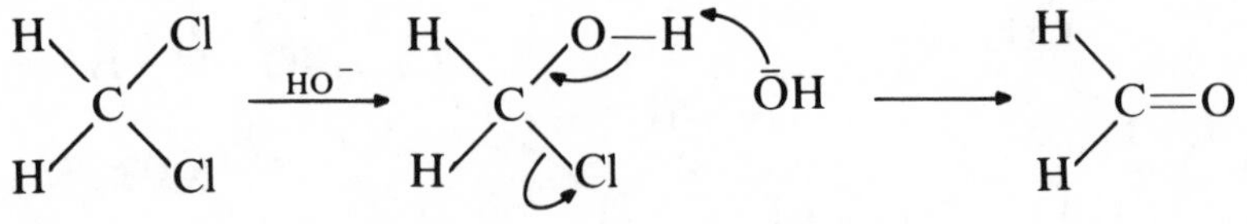

Other dihalogenoalkanes react similarly. Carbon tetrachloride (tetrachloromethane), CCl_4, is very unreactive towards nucleophiles, as are other polyhalogenoalkanes, e.g. Teflon, mentioned in Chapter 5.

Chloroform (trichloromethane), $CHCl_3$, reacts with bases, but in a rather different reaction. The presence of three electronegative chlorine atoms weakens the carbon–hydrogen bond, and an anion, $[CCl_3]^-$, is formed:

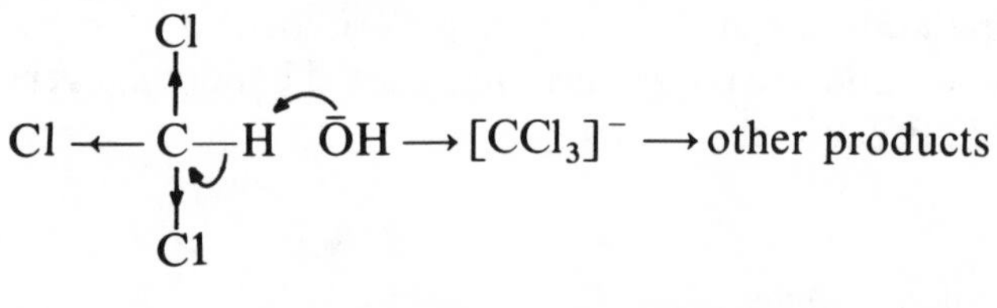

This anion itself readily decomposes giving a number of products.

Uses of Polyhalogeno Compounds

Chloroform was once used extensively as an anaesthetic, but it is now known to have deleterious effects on many organs of the body and is consequently no longer used. A popular inhalative anaesthetic is another polyhalogeno compound, called 'Halothane', CF_3—$CHClBr$. This is relatively safe and non-toxic and, like all polyhalogeno compounds, has the advantage of being non-inflammable.

The non-inflammability of these compounds has led to their extensive use as solvents and also as fire-extinguishers. Carbon tetrachloride was commonly used for both of these purposes, in commercial fire-extinguishers and as a dry-cleaning solvent. However, as its vapours cause severe damage to the liver, it is no longer in favour for these purposes. It also presented a danger in fire-fighting in that at the temperature of the fire it reacted with oxygen to form phosgene, $COCl_2$, which was used as a poison gas in the 1914–1918 World War. Operatives who smoked while using carbon tetrachloride in dry-cleaning were at risk in that the burning tobacco also oxidized the vapour of the solvent to generate phosgene.

Chlorofluorocarbons such as CCl_3F or CCl_2F_2 have been used extensively as refrigerants in refrigerators and as solvents in aerosol sprays. They are gases at normal temperatures but readily liquefy when compressed. In the cooling coils of refrigerators they evaporate and expand, absorbing heat in the process and thereby cooling their surroundings. The evaporated gas is then compressed again and available to repeat the process. When an aerosol is operated the solution is forced through the nozzle of the can and the volatile solvent at once evaporates leaving a fine spray of the material contained in it. However, there is now fear that these chlorofluorocarbons, of which thousands of tons are discharged into the atmosphere every year, may have deleterious effects on the upper atmosphere and in consequence their use is being curtailed.

Finally mention may be made again of the very resistant fluorocarbon polymers already mentioned in Chapter 5.

Detection of Halogens in Organic Compounds

The classical method of detection of the presence of halogens in organic molecules used the **Lassaigne test**, in which a very small sample of the material under test was fused with molten sodium (for details see textbooks of practical organic chemistry). In the process the halogen atoms were converted into halide ions, which were then tested for by the standard method using silver nitrate.

The presence of chlorine and of bromine may readily be shown by mass spectrometry (for a description of this technique, see the Appendix). This provides accurate measures of the molecular masses of molecules. Chlorine exists naturally as a 3:1 ratio of ^{35}Cl and ^{37}Cl isotopes. Hence a compound containing one atom of chlorine in its molecules shows two peaks separated by two mass units, the peak at lower mass being three times the height of the other. Thus methyl chloride, CH_3Cl, shows peaks at 50 and 52, in the ratio 3:1. Bromine is a mixture of approximately equal amount of the isotopes ^{79}Br and

^{81}Br. Thus a compound containing one atom of bromine in its molecule also shows two peaks separated by two mass units, but in this case the two peaks will be of almost equal height. Methyl bromide, CH_3Br, shows peaks at 94 and 96. Note that a dibromo compound will provide **three** peaks, each separated by two mass units, with relative heights 1:2:1. Iodine has only one isotope, ^{127}I, but in the mass spectrometer an iodine is readily lost from the molecule, as I^+, and the mass spectrum of an iodo compound usually shows an abundant peak at 127, due to I^+.

In Conclusion—Might Other Compounds Resemble Alkyl Halides?

What other compounds might have dipolar bonds like alkyl halides? Elements at the top right-hand corner of the standard Periodic Table are electronegative and hence should form polar bonds with carbon. Thus carbon–oxygen and carbon–nitrogen bonds might be expected to be polarized.

$$R \longrightarrow Cl \qquad R \longrightarrow O— \qquad R \longrightarrow N\big\langle$$

Experience shows that halide ions form better leaving groups than do anions involving oxygen, such as hydroxide, which in turn are better leaving groups than anions containing nitrogen, such as H_2N^- or R_2N^-.

Thus we should expect reactivity towards nucleophiles to decrease in the order $RCl > ROH > RNH_2$. As will be described in the following chapters, this is the case, Thus:

(a) alkyl halides, RCl, RBr, RI, react readily;
(b) alcohols, ROH, and ethers, ROR, react only in the presence of acid;
(c) amines, RNH_2, do not react with nucleophiles to lose H_2N^-.

This demonstrates that the effectiveness as a leaving group decreases in the order $Cl^- > HO^- > H_2N^-$.

This character is also reflected in the acidities of the related compounds wherein a hydrogen atom replaces an alkyl group. The compounds in question are hydrogen chloride, water and ammonia.

Hydrogen chloride is a moderately strong acid and readily loses Cl^-:

$$H—Cl \xrightarrow{H_2O} H_3O^+ + Cl^-$$

Water is a very weak acid, not readily losing HO^-:

$$H-OH \underset{H_2O}{\rightleftharpoons} H_3O^+ + HO^-$$

Ammonia does not ionize in water by loss of a proton, although it will form salts with metals:

$$2\,NH_3 + 2\,Na \longrightarrow 2\,Na^+ \quad 2\,H_2N^- + H_2\uparrow$$
$$(\text{cf. } 2\,H_2O + 2\,Na \longrightarrow 2\,Na^+ \quad 2\,HO^- + H_2\uparrow)$$

Questions

1. What are the structural formulae of (a) methyl chloride, (b) 2-bromobutane, (c) iodocyclopentane and (d) 3-ethyl-1-fluoro-2-methylhexane?
2. Distinguish between homolytic and heterolytic cleavage of a chemical bond. Which type of reaction is involved in (a) the reaction of an alkane with chlorine and (b) the reaction of an alkyl halide with a nucleophile?
3. What are the reaction products from the following reactions? Give general formulae and the names of the types of compound that are formed:
   (a) $RBr + HO^-$, (b) $RBr + RO^-$, (c) $RCl + NH_3$.
4. In the reaction $NaI + RCl \rightleftharpoons RI + NaCl$ how can the formation of the alkyl iodide be assisted?
5. If you wished to prepare $CH_3OC(CH_3)_3$, which of the following reactions would you use:
   (a) $CH_3Cl + (CH_3)_3C-O^-Na^+$ or (b) $(CH_3)_3CCl + CH_3O^-Na^+$?
   What organic products might you expect to obtain from each of these reactions?
6. Does the following reaction take place?
   $2\,C_2H_5Cl + Na_2SO_4 \longrightarrow (C_2H_5)_2SO_4$
   Explain the reason for your answer.
7. How many signals would you expect to see in the ^{13}C-n.m.r. spectra of (a) chloroethane, (b) 1,2-dichloroethane, (c) 1,1-dichloroethane, (d) 1-chloropropane and (e) 2-chloropropane?
8. Complete the following reaction sequence:

$$CH_2{=}CH_2 \xrightarrow{?} ? \xrightarrow{?} CH_3CH_2CN$$

9. A compound (X) has the molecular formula C_4H_9Cl and shows one signal in its ^{1}H-n.m.r. spectrum and two signals in its ^{13}C-n.m.r. spectrum. When treated with base it is converted into another compound (Y), which has the molecular formula C_4H_8 and shows two signals in its ^{1}H-n.m.r. spectrum and three signals in its ^{13}C-n.m.r. spectrum.

    What are the structural formulae of X and Y? Write a possible mechanism for the conversion of X into Y.

8 ALKYL AMINES

Amines are organic derivatives of ammonia. Alkyl amines have one or more alkyl groups attached to a nitrogen atom, as in RNH_2, R_2NH and R_3N. Because in these molecules one, two or three hydrogen atoms of ammonia are replaced by alkyl groups, they are known, respectively, as **primary, secondary** and **tertiary amines**.

Ammonia is a gas that is very soluble in water because there is hydrogen bonding between the ammonia and water molecules. Amines of low molecular weight are gases, or liquids of low boiling point; as the molecular weights get larger so do the boiling and melting points. Alkyl amines of low molecular weight are soluble in water, again because of hydrogen bonding (see page 98) between water and amine molecules:

$$R_3N: \cdots\cdots\cdots \overset{\delta+}{H} - \overset{\delta-}{O} - \overset{\delta+}{H}$$

As well as amines with separate alkyl groups attached to nitrogen, as in CH_3NH_2, $(CH_3)_2NH$ and $(CH_3)_3N$, there are amines in which the nitrogen atom forms part of a ring, as in

These are secondary amines if the nitrogen atom has a hydrogen atom

attached to it or tertiary amines if that hydrogen atom is replaced by an
alkyl group, as in

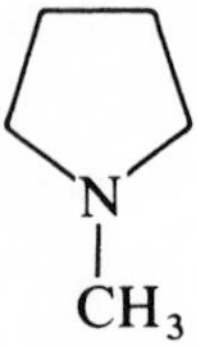

Their properties closely resemble those of non-cyclic (or **acyclic**)
secondary and tertiary amines.

Nomenclature of Amines

Amines are named by adding the names of the alkyl groups attached to
the nitrogen atom as prefixes to the word amine, for example

CH_3NH_2	Methylamine
$(CH_3)_2NH$	Dimethylamine
$(CH_3)_3N$	Trimethylamine
$CH_3NHC_2H_5$	Ethylmethylamine

Alternatively they are named as alkanes with the prefix **amino** to denote
the attached group, as in

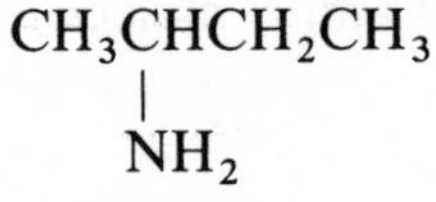

$CH_3CHCH_2CH_3$
$\quad\;\;|$
$\quad NH_2$

$\qquad\qquad\qquad CH_3\quad CH_3$
$\qquad\qquad\qquad\;\;\diagdown N \diagup$
$\qquad\qquad\qquad\qquad |$
$\qquad\qquad\quad CH_3CH_2CHCH_2CH_3$

2-Aminobutane 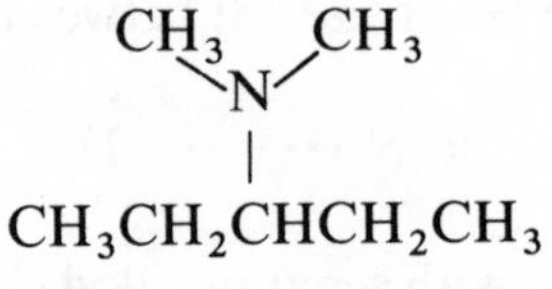   3-(Dimethylamino)pentane

$\qquad\qquad NH_2CH_2CH_2NH_2 \quad$ 1, 2-Diaminoethane

The latter method is advantageous if the amino group is not attached to
the end of a carbon chain.

The cyclic amines mentioned above have official names which are
non-systematic, namely

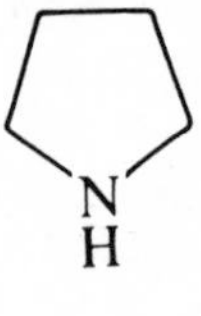

Pyrrolidine

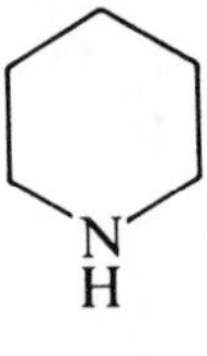

Piperidine

Isomerism of Amines

Inevitably there are many isomeric amines. To take a simple example, the molecular formula C_3H_9N may represent four isomers:

$(CH_3)_3N$	Trimethylamine	(a tertiary amine)
$CH_3NHCH_2CH_3$	Ethylmethylamine	(a secondary amine)
$CH_3CH_2CH_2NH_2$	Propylamine or	(a primary amine)
	1-aminopropane	
CH_3CHCH_3 $\quad\vert$ $\quad NH_2$	2-Aminopropane	(a primary amine)

Of these isomers two are isomeric primary amines, one is a secondary amine and one a tertiary amine.

They can be distinguished from each other by their n.m.r. spectra:

1. In $(CH_3)_3N$ all the hydrogen atoms are equivalent and so are all the carbon atoms. Therefore the ^1H-n.m.r. spectrum and the ^{13}C-n.m.r. spectrum each consist of only one peak.
2. In $CH_3CH_2NHCH_3$ there are two different methyl groups, one attached to a CH_2 group and one to the NH group. Thus all three carbon atoms are in different environments and there will be three signals in the ^{13}C-n.m.r. spectrum. The hydrogen atoms are in four different environments (two different CH_3 groups, one CH_2 group, one NH group) and so will provide four separate signals, the ratios of the numbers of hydrogen atoms in these environments, obtainable from the ^1H-n.m.r. spectrum, being 3:3:2:1.
3. Similar arguments may be applied to the other isomers. $CH_3CH_2CH_2NH_2$ produces three signals in its ^{13}C-n.m.r. spectrum (CH_3, CH_2 and CH_2) and four in its ^1H-n.m.r. spectrum with a ratio of numbers of hydrogen atoms $= 3:2:2:2$.

4. $\quad CH_3CHCH_3$
$\qquad\vert$
$\qquad NH_2$ provides two carbon signals (two equivalent methyl groups and one CH group) and three hydrogen signals ($2CH_3$, CH and NH_2) in the ratio 6:1:2.

Basicity of Amines

When ammonia dissolves in water it provides an alkaline solution because of the equilibrium which generates hydroxide ions:

$$NH_3 + H_2O \rightleftharpoons \overset{+}{N}H_4 + HO^-$$

In terms of curved arrows the reaction may be expressed as follows:

$$H_3N: \frown H \frown OH \rightleftharpoons H_3\overset{+}{N} - H \frown {}^-OH$$

As mentioned at the end of the previous chapter, ammonia is a stronger base than water and hence an ammonia molecule may remove a proton from a water molecule.

Alkyl amines can react similarly with water, for example

$$CH_3NH_2 + H_2O \rightleftharpoons CH_3\overset{+}{N}H_3 + HO^-$$

They are somewhat stronger bases than ammonia.

Basicity is the most characteristic property of alkyl amines. Like ammonia, they readily form salts with acids, e.g. with hydrogen chloride:

$$NH_3 + HCl \longrightarrow \overset{+}{N}H_4Cl^-$$ Ammonium chloride

$$CH_3NH_2 + HCl \longrightarrow CH_3\overset{+}{N}H_3Cl^-$$ Methylammonium chloride

$$(CH_3)_2NH + HCl \longrightarrow (CH_3)_2\overset{+}{N}H_2Cl^-$$ Dimethylammonium chloride

$$(CH_3)_3N + HCl \longrightarrow (CH_3)_3\overset{+}{N}HCl^-$$ Trimethylammonium chloride

Other acids react similarly.

Alkyl Amines as Nucleophiles

As well as being a base, ammonia is also a strong nucleophile. As discussed in the previous chapter, ammonia reacts with alkyl halides to form alkylammonium salts, for example

$$H_3N: \frown CH_3 \frown Br \longrightarrow H_3\overset{+}{N}CH_3 \; Br^-$$ Methylammonium bromide (a primary ammonium salt)

Amines react similarly:

Primary amine $RNH_2 + R'Br \longrightarrow R\overset{+}{N}H_2R' \quad Br^-$

Secondary ammonium salt

Secondary amine $R_2NH + R'Br \longrightarrow R_2\overset{+}{N}HR' \quad Br^-$

Tertiary ammonium salt

Tertiary amine $R_3N + R'Br \longrightarrow R_3\overset{+}{N}R' \quad Br^-$

Quaternary ammonium salt

When they react with alkyl halides, ammonia, primary amines and secondary amines give, respectively, primary ammonium salts, secondary ammonium salts and tertiary ammonium salts. Tertiary amines give **quaternary ammonium salts**, in which all four of the hydrogen atoms which would be attached to nitrogen in $\overset{+}{N}H_4$ have been replaced by alkyl groups.

The situation is not quite so simple in practice and mixtures of different types of ammonium salts are commonly obtained. Consider the reaction of ammonia with methyl iodide, which first of all provides methylammonium iodide:

$$NH_3 + CH_3I \longrightarrow CH_3\overset{+}{N}H_3 \quad I^-$$

Any unreacted ammonia present may react with this salt to generate some methylamine:

$$CH_3\overset{+}{N}H_3 I^- + NH_3 \rightleftharpoons CH_3NH_2 + \overset{+}{N}H_4 \quad I^-$$

This methylamine may now react with more methyl iodide to give dimethylammonium iodide:

$$CH_3NH_2 + CH_3I \longrightarrow (CH_3)_2\overset{+}{N}H_2 \quad I^-$$

By a similar series of reactions trimethylammonium and tetramethyl-ammonium iodides may be formed. Thus a mixture of methyl-, dimethyl-, trimethyl- and tetramethyl-ammonium salts can result from the reaction of ammonia with methyl iodide. The proportions of these different salts in the final product may depend on a number of factors, including, of course, the ratios of the amounts of the original reactants, ammonia and methyl iodide. Hence, as is often the case in organic chemistry, mixtures of products may arise in carrying out reactions, and perhaps **the** major skill in organic chemistry lies in separating out such mixtures.

Action of Alkali on Alkylammonium Salts

Tertiary, secondary and primary alkylammonium salts generate, respectively, tertiary, secondary and primary amines when they are treated with alkali. The latter removes a proton from the nitrogen atom:

$$R-\overset{\overset{\displaystyle R}{|}}{\underset{\underset{\displaystyle R}{|}}{\overset{+}{N}}}-H \quad {}^-OH \longrightarrow R_3N + H_2O$$

$$R-\overset{\overset{\displaystyle H}{|}}{\underset{\underset{\displaystyle R}{|}}{\overset{+}{N}}}-H \quad {}^-OH \longrightarrow R_2NH + H_2O$$

$$R-\overset{\overset{\displaystyle H}{|}}{\underset{\underset{\displaystyle H}{|}}{\overset{+}{N}}}-H \quad {}^-OH \longrightarrow RNH_2 + H_2O$$

However, quaternary ammonium salts cannot take part in such reactions because they do not have hydrogen atoms directly attached to the nitrogen atom.

In the cold, all that happens is that the salt is transformed into a quaternary ammonium hydroxide:

$$R_4\overset{+}{N} \quad Cl^- \xrightarrow{\text{HO}^-} R_4N^+ \quad HO^-$$

These quaternary ammonium hydroxides are very strong bases.

When heated with alkali, quaternary salts which have hydrogen atoms on a β-carbon atom undergo an elimination reaction to give a tertiary amine and an alkene, for example

$$(CH_3)_3\overset{+}{N}C_2H_5 \quad HO^- \xrightarrow{\text{heat}} (CH_3)_3N + CH_2{=}CH_2 + H_2O$$

The mechanism resembles that of elimination reactions of alkyl halides, namely

$$CH_3-\overset{\overset{\displaystyle CH_3}{|}}{\underset{\underset{\displaystyle CH_3}{|}}{N^+}}-CH_2-\overset{\overset{\displaystyle H}{|}}{\underset{\underset{\displaystyle H}{|}}{C}}-H \qquad \overset{^-OH}{} \qquad \xrightarrow{\text{heat}} \quad CH_3-\overset{\overset{\displaystyle CH_3}{|}}{\underset{\underset{\displaystyle CH_3}{|}}{N}}: \; + CH_2{=}CH_2 + H_2O$$

Tertiary, secondary and primary ammonium salts do not undergo this reaction even if they have β-carbon atoms with hydrogen atoms attached to them. The base preferentially attacks a hydrogen atom attached to the nitrogen atom, for example

$$CH_3-\overset{\overset{\displaystyle H}{|}}{\underset{\underset{\displaystyle CH_3}{|}}{N}}-C_2H_5 \leftarrow \;\; HO^- \quad H \quad CH_3-\overset{\overset{\displaystyle H}{|}}{\underset{\underset{\displaystyle CH_3}{|}}{N^+}}-CH_2-\overset{\overset{\displaystyle H}{|}}{\underset{\underset{\displaystyle H}{|}}{C}}-H \quad ^-OH$$

Questions

1. Four isomers of C_3H_9N have ^{1}H-n.m.r. spectra as follows:
   (a) Four signals in ratio 3:2:2:2
   (b) Three signals in ratio 6:2:1
   (c) Four signals in ratio 3:3:2:1
   (d) One signal only

   Write the structural formulae of these isomers.

2. What organic products are obtained if the following salts are treated with potassium hydroxide:
   (a) $(C_2H_5)_3\overset{+}{N}H\,Cl^-$,
   (b) $(C_2H_5)_4\overset{+}{N}\,Cl^-$,
   (c) ethyldimethylammonium chloride and
   (d) ethyltrimethylammonium chloride?

9 ALCOHOLS

As alkyl amines are organic derivatives of ammonia, so alcohols and ethers are organic derivatives of water, in which, respectively, one and two hydrogen atoms are replaced by alkyl groups:

H—O—H   R—O—H   R—O—R (or R—O—R′)
Water   Alcohols   Ethers

This chapter considers alcohols; ethers are dealt with in the next chapter. The characteristic **functional group** of an alcohol, the —O—H group, is known as a **hydroxy** group.

Nomenclature

Alcohols are named by adding the ending **ol** to the name of the hydrocarbon to which they are related; for example

CH_3OH CH_3CH_2OH $CH_3CHOHCH_3$
Methanol   Ethanol   Propan-2-ol*   Cyclohexanol

*In the United States 2-propanol is used; see a footnote in Chapter 3 (p. 30) about the placing of numerals in names.

Common Alcohols

Methanol and ethanol are both common chemicals, made industrially on a very large scale.

Methanol is made on a huge scale industrially from a mixture of carbon monoxide and hydrogen which is heated under high pressure in the presence of a catalyst:

$$CO + 2H_2 \xrightarrow[\text{catalyst}]{\text{heat, pressure,}} CH_3OH$$

This mixture of carbon monoxide and hydrogen is obtained from the partial oxidation of methane, which is itself obtained either from natural gas, the cracking of hydrocarbons (see Chapter 2) or by gasification of coal. Natural gas and oil are the commonest present source but as supplies of these materials run out, coal will take over. Until about the mid-1920s methanol was made by dry distillation of wood. Methanol is of very great importance as an intermediate from which other organic compounds are made. For example, it is oxidized to formaldehyde (see Chapter 14) and reacts with carbon monoxide in the presence of certain catalysts to give acetic acid (see Chapter 15).

Methanol is extremely toxic. In small doses it causes abdominal pain, nausea, muscular weakness and defective heart action. Larger amounts lead to blindness, coma and death.

Ethanol is a product from the fermentation of fruit and grains and other naturally occurring sources of sugar, and has been made in dilute aqueous solution from these sources since time immemorial. Thus, fermentation of apples, grapes and barley has provided, respectively, cider, wine and beer. It is possible that fermentation of natural materials may in the future provide a valuable source of ethanol for industrial purposes; some is already being produced in this way.

Most industrial ethanol, however, is obtained from ethylene, as discussed in Chapter 5.

Ethanol is another industrial chemical of enormous importance, for use as an intermediate from which other compounds are made and also as a solvent.

Ethanol is a normal metabolite. Although less toxic than methanol, it is toxic in large quantities.

Isomerism in Alcohols

As with all classes of organic compounds, isomerism is possible. This may be exemplified by considering the possible alcohols with molecular formula C_4H_9OH. They are

(a) $CH_3CH_2CH_2CH_2OH$ Butan-1-ol
(b) $CH_3CH_2CHOHCH_3$ Butan-2-ol
(c) CH_3CHCH_2OH 2-Methylpropan-1-ol
     |
    CH_3

     CH_3
     |
(d) CH_3C—OH   2-Methylpropan-2-ol
     |
    CH_3

These alcohols are sometimes known by their older non-systematic names which are, respectively, (a) n-butanol, (b), secondary (or *sec-*) butanol, (c), isobutanol and (d) tertiary (or t-) butanol.

They can most easily be distinguished from one another by their n.m.r. spectra. Thus (a), (b), (c) and (d) show, respectively, 4, 4, 3 and 2 signals in their ^{13}C-n.m.r. spectra. Alcohols (a) and (b) each have hydrogen atoms in five different environments, but the ratios of the numbers of hydrogen atoms in the different environments is different for the two alcohols. In practice, but not to be discussed here, the splitting patterns (see the Appendix) in the ^{1}H-n.m.r. spectra of (a) and (b) would be different and distinguish between the two.

Alcohols are also isomeric with ethers. For example, the simple molecular formula C_2H_6O represents both the alcohol ethanol, CH_3CH_2OH, and an ether, CH_3OCH_3. N.m.r. spectra would again easily distinguish between these two compounds, ethanol showing two signals in its ^{13}C-n.m.r. spectrum and three signals in its ^{1}H-n.m.r. spectrum, while the ether would give only one signal in both its ^{13}C-n.m.r. and ^{1}H-n.m.r. spectra.

Another easy way of distinguishing between alcohols and ethers is by using infrared spectroscopy. The hydroxy group of alcohols provides a signal at about 3250–3650 cm^{-1} in their i.r. spectra. Since ethers have no hydroxy group there is no such signal in their i.r. spectra.

(Unless you were in a very cold place it would be rather unnecessary to use spectroscopy to distinguish between ethanol, C_2H_5OH, and methyl ether, CH_3OCH_3, since at 'normal' temperatures, ethanol is a liquid, b.p. 78 °C, and methyl ether is a gas, b.p. -23 °C. There is no point in using complicated techniques if simpler ones suffice!)

Physical Properties

It has been seen in the case of alkanes that boiling points are in general higher the greater the molecular weight. Yet, as described in the previous paragraph, ethanol and methyl ether, although having the same molecular formula and molecular weight, have very different boiling points. This difference is associated with the presence of a hydroxy group in ethanol but not in methyl ether.

Water has a remarkably high boiling point if it is considered as H_2O, with a molecular weight of 18. However, water molecules are closely **associated** with one another by so-called **hydrogen bonding**:

$$
\begin{array}{ccccc}
H & & H & & H \\
\diagdown & & \diagdown & & \diagdown \\
\cdots\cdots O-H & \cdots\cdots & O-H & \cdots\cdots & O-H & \cdots\cdots
\end{array}
$$

Thus water does not exist as isolated H_2O molecules, but all the molecules interact with one another. This happens because O—H bonds are polarized: O→H. Hence the oxygen atoms bear a small negative charge and the hydrogen atoms a small positive charge:

$$
\overset{\delta+}{H}-\overset{\delta-}{O}-\overset{\delta+}{H}
$$

This leads to electrostatic interaction between adjacent molecules, between the oxygen atom of one molecule and a hydrogen atom of another molecule:

$$
\begin{array}{ccccc}
H & & H & & \\
\diagdown & & \diagdown & & \\
\text{etc}\cdots\cdots \underset{\delta-}{O}-\underset{\delta+}{H} & \cdots\cdots & \underset{\delta-}{O}-\underset{\delta+}{H} & \cdots\cdots \text{etc.}
\end{array}
$$

A similar interaction takes place between alcohol molecules:

$$
\begin{array}{ccccc}
R & & R & & R \\
\diagdown & & \diagdown & & \diagdown \\
\cdots\cdots \underset{\delta-}{O}-\underset{\delta+}{H} & \cdots\cdots & \underset{\delta-}{O}-\underset{\delta+}{H} & \cdots\cdots & \underset{\delta-}{O}-\underset{\delta+}{H}
\end{array}
$$

In contrast ethers have no hydroxy groups and consequently hydrogen bonding is not possible. This in turn means that there is no such strong association between different ether molecules, and this makes their boiling points much lower than those of isomeric alcohols.

As usual, the boiling points of alcohols increase with increasing molecular weight, e.g. for methanol, b.p. $= 65\,^\circ$C, for ethanol, b.p. $= 78\,^\circ$C, for butan-1-ol, b.p. $= 117\,^\circ$C.

Solubility in Water

In the same way that water molecules can interact with one another by hydrogen bonding, and ethanol molecules can do likewise, so water and ethanol molecules can interact with one another:

$$\cdots\cdots \overset{\overset{\textstyle H}{\diagdown}}{O}-H\cdots\cdots\overset{\overset{\textstyle C_2H_5}{\diagdown}}{O}-H\cdots\cdots\overset{\overset{\textstyle H}{\diagdown}}{O}-H\cdots\cdots\overset{\overset{\textstyle C_2H_5}{\diagdown}}{O}-H\cdots\cdots$$

A consequence of this is that ethanol dissolves freely in water. As the size of the alkyl group in an alcohol increases, the solubility of the alcohol in water decreases. Whereas methanol, ethanol and propan-1-ol are infinitely soluble in water, only nine parts of butan-1-ol and only three parts of pentan-1-ol will dissolve in 100 parts of water. Alkanes are insoluble in water, and as the size of the alkyl group increases so its solubility properties hide and override those of the hydroxy group.

Amphoteric Character of Alcohols

As derivatives of water, alcohols have some properties related to those of water, namely those associated with the hydroxy group.

Perhaps the most characteristic property of water is that it is amphoteric:

$$2\,H_2O \rightleftharpoons H_3O^+ + {}^-OH$$
$$1 \quad : \quad 10^{-7} \quad : \quad 10^{-7} \qquad \text{(Ratio of amounts present)}$$

$$H_2O \begin{array}{l} \xrightarrow{\;\overset{acid}{H^+\,X^-}\;} H_3O^+ + X^- \quad (\text{e.g.} + HCl \longrightarrow H_3O^+ + Cl^-) \\[2mm] \xrightarrow{\;\underset{B:}{base}\;} BH^+ + {}^-OH \quad (\text{e.g.} + NH_3 \longrightarrow \overset{+}{N}H_4 + {}^-OH) \\[2mm] \xrightarrow{\;\underset{B^-}{base}\;} BH + {}^-OH \quad (\text{e.g.} + {}^-NH_2 \longrightarrow NH_3 + {}^-OH) \end{array}$$

Another example of the acidity of water is its reaction with sodium to give sodium hydroxide:

$$2\,H_2O \xrightarrow{2Na} H_2\uparrow + 2\,Na^+ \quad {}^-OH$$

All these reactions are also characteristic of alcohols. For example

$$2\,CH_3OH \rightleftharpoons CH_3\overset{+}{O}H_2 + {}^-OCH_3$$
$$1 \quad : \quad 10^{-8} \quad : -10^{-9} \text{ (Ratio of amounts present)}$$

The degree of dissociation of methanol is even less than in the case of water. Alcohols of higher molecular weight are even less dissociated. Alcohols also react with acids and with bases:

$$ROH \underset{B:}{\overset{H^+X^-}{\bigg\langle}} \begin{array}{l} R\overset{+}{O}H_2 + X^- \\ \\ BH^+ + {}^-OR \end{array}$$

Again alcohols are weaker acids and weaker bases than water.

Alcohols also react with sodium (or other alkali metals) to give hydrogen and **alkoxide** ions:

$$2\,ROH \xrightarrow{2Na} H_2\uparrow + 2\,Na^+ \quad {}^-OR$$

It was mentioned that alcohols are slightly weaker acids than water. Conversely the alkoxide ions are stronger bases than the hydroxide ion. This can be represented by the equilibrium:

$$C_2H_5O^- \quad Na^+ + H_2O \rightleftharpoons C_2H_5OH + Na^+ \quad {}^-OH$$

Hence sodium hydroxide does not generate any appreciable amount of sodium ethoxide when it dissolves in ethanol. To prepare sodium alkoxides it is necessary to dissolve sodium in the alcohol. The alkoxides are very hygroscopic and react readily with water to give the corresponding alcohol:

$$RO^- + H_2O \rightleftharpoons ROH + HO^-$$

Reactions of Alcohols with Halogen Acids

It was mentioned at the end of Chapter 7 that alcohols do not react with nucleophiles except in the presence of an acid. For example, ethanol does not react with potassium iodide, but with hydrogen iodide it reacts to give iodoethane:

$$C_2H_5OH \overset{KI}{\underset{HI}{\Big\langle}} \begin{array}{l} \text{no reaction} \\ \\ C_2H_5I \end{array}$$

In the reaction with the acid the alcohol first of all acts as a base and is protonated. The protonated alcohol then reacts with iodide ion, in a nucleophilic substitution reaction, to give iodoethane:

$$C_2H_5OH + HI \longrightarrow C_2H_5\overset{+}{O}H_2 \quad I^-$$

$$I^- \quad \overset{\frown}{CH_2} \overset{+}{\underset{CH_3}{-}OH_2} \longrightarrow CH_3CH_2I + H_2O$$

Two factors conspire to make the protonated alcohol react where the unprotonated alcohol does not. The positively charged oxygen atom, which is called an **oxonium ion**, provides a greater inductive effect in the bond linking it to the adjacent carbon atom than does a hydroxy group. In addition, water is a better leaving group than is a hydroxide ion.

Bromoethane may be made similarly from ethanol. A mixture of sulphuric acid and potassium bromide is often used instead of hydrogen bromide:

$$C_2H_5OH \xrightarrow[KBr]{H_2SO_4} C_2H_5Br$$

The corresponding reaction with hydrogen chloride is less successful. Chloride ion is a poorer nucleophile than bromide or iodide ions. Reaction takes place satisfactorily with a tertiary alcohol, i.e. an alcohol in which the hydroxy group is attached to a tertiary carbon atom, but not with a primary alcohol, whose hydroxy group is attached to a primary carbon atom.

Conversion of an alcohol into an alkyl chloride is commonly done by reaction with either thionyl chloride or phosphorus pentachloride:

$$ROH + SOCl_2 \longrightarrow RCl + SO_2 + Cl_2$$
$$ROH + PCl_5 \longrightarrow RCl + HCl + POCl_3$$

The mechanisms of these reactions are more complex and will not be discussed here. The method using thionyl chloride is the more useful for the preparation of chloro compounds because it gives a cleaner product. The other products are gases and boil out of the solution, whereas when phosphorus pentachloride is used, a residue remains from which the alkyl chloride has to be separated.

Tests for Alcohols

Although less satisfactory as a preparative method, the reaction of an alcohol with phosphorus pentachloride has been used as a test for alcohols since reaction is usually fierce and easily observed, accompanied by the evolution of hydrogen chloride. The classical chemical tests to show the presence of an alcohol have been this reaction with phosphorus pentachloride and the reaction with sodium, which dissolves and hydrogen is evolved.

Nowadays recourse is usually made to an infrared spectrum, and the presence or absence of a characteristic absorption at about 3250–3650 cm^{-1}, which is associated with the presence of a hydroxy group.

To be reliable, both the chemical and spectroscopic tests require that the sample under test is dry, for any water present as an impurity would give misleading positive results, since it too contains a hydroxy group.

Reactions of Alcohols with Nucleophiles

As has just been seen, halide ions only react with alcohols in the presence of acid.

It is not possible, therefore, for basic nucleophiles such as ammonia or cyanide ion to react with alcohols to give amines or cyanides, because, under the acidic conditions required, the nucleophiles themselves are protonated and the resultant products are no longer nucleophiles:

$$NH_3 + H^+ \longrightarrow \overset{+}{N}H_4$$
$$CN^- + H^+ \longrightarrow HCN$$

Reactions of Alcohols with Acids

In the reactions considered above of alcohols with halogen acids the products are alkyl halides, for example

$$ROH + HI \longrightarrow RI + H_2O$$

In general the reactions of alcohols with acids may be expressed as follows:

$$ROH + HX \rightleftharpoons RX + H_2O$$
$$Alcohol + Acid \rightleftharpoons Ester + Water$$

The general name for the product is an **ester**; an ester may be defined as a compound in which the acidic hydrogen of an acid has been replaced by an alkyl group. (Strictly, from this definition, alkyl halides can be thought of as esters derived from halogen acids, but conventionally, and because there are great practical advantages in doing so, these compounds are classed separately from esters.)

Note that when alcohols react with an acid conversion into an ester and water is incomplete and an equilibrium is set up, as shown above. This will be considered in more detail later.

Organic as well as inorganic acids may participate in this type of reaction, for example

$$CH_3COOH + C_2H_5OH \rightleftharpoons CH_3COOC_2H_5 + H_2O$$
$$Acetic\ acid \quad Ethanol \quad Ethyl\ acetate$$

An example involving an inorganic acid is as follows:

$$ROH + HONO_2 \rightleftharpoons RONO_2 + H_2O$$
$$Nitric \quad An\ alkyl$$
$$acid \quad nitrate$$

Many alkyl nitrates are explosive.

An **esterification** reaction which is important biologically is that involving phosphoric acid. Since phosphoric acid has three acidic hydrogen atoms, mono-, di- and tri-esters may be formed:

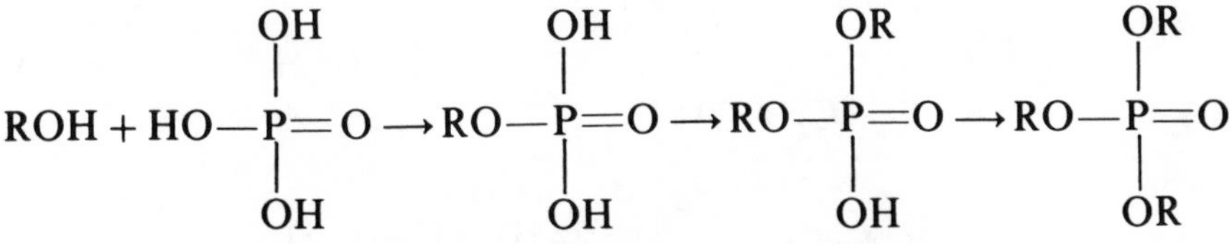

$$
\begin{array}{ccccc}
& OH & OH & OR & OR \\
& | & | & | & | \\
ROH + HO{-}P{=}O & \rightarrow RO{-}P{=}O & \rightarrow RO{-}P{=}O & \rightarrow RO{-}P{=}O \\
& | & | & | & | \\
& OH & OH & OH & OR
\end{array}
$$

Phosphoric acid	A monoalkyl phosphate	A dialkyl phosphate	A trialkyl phosphate

Complications due to competing reactions may sometimes arise.

Let us consider the reaction of ethanol with sulphuric acid. In the cold, using molar equivalents of the reagents, a monoester is formed:

$$1\,C_2H_5OH + 1\,HOSO_2OH \underset{\text{cold}}{\rightleftharpoons} C_2H_5OSO_2OH + H_2O$$

Sulphuric acid   Ethyl hydrogen
                     sulphate (or ethyl sulphate)

If more ethanol is present, some diethyl sulphate, $C_2H_5OSO_2OC_2H_5$, will be formed.

These sulphate esters behave chemically like alkyl halides, but with an even better leaving group than a halide ion because the sulphate anion, being derived from a stronger acid than the halogen acids, is an extremely effective leaving group. Because sulphate is such a good leaving group, a sulphate ester may go on to take part in further reactions under suitable circumstances.

As an example of this, if ethyl sulphate is heated with excess ethanol, the sulphate group is displaced by nucleophilic attack by ethanol, and an ether is formed:

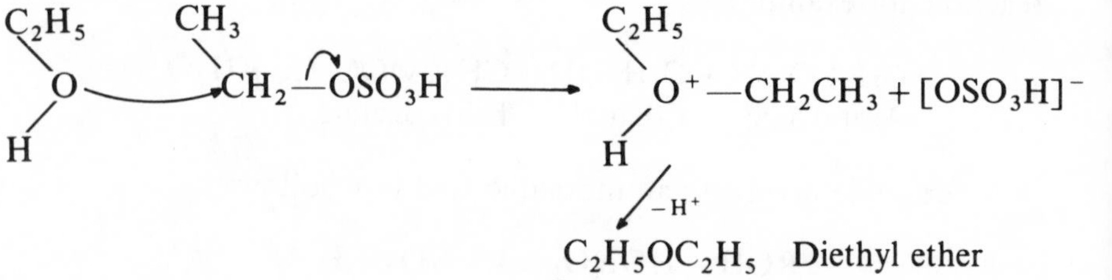

Diethyl ether can indeed be made from the reaction of sulphuric acid with an excess of ethanol, but the ether may also be formed by an alternative mechanism, involving initial protonation of ethanol to give an oxonium ion, which is then attacked by another molecule of ethanol:

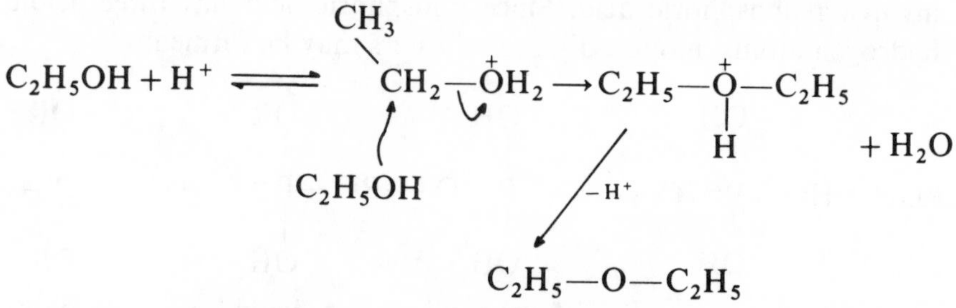

In this case water acts as an excellent leaving group.

As compounds having good leaving groups, alkyl sulphates can also undergo elimination reactions. For example, if ethyl sulphate is heated, ethylene is formed:

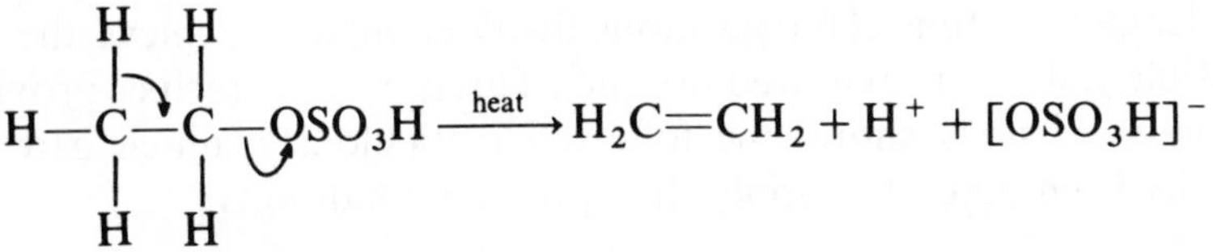

$$H-\overset{\overset{\displaystyle H}{|}}{\underset{\underset{\displaystyle H}{|}}{C}}-\overset{\overset{\displaystyle H}{|}}{\underset{\underset{\displaystyle H}{|}}{C}}-OSO_3H \xrightarrow{\text{heat}} H_2C{=}CH_2 + H^+ + [OSO_3H]^-$$

In the same way that, in the formation of diethyl ether from sulphuric acid and excess ethanol, it is not necessary for the ethyl sulphate to be formed first, so it need also not be an intermediate in the formation of ethylene. Once again protonated ethanol could perform the same role:

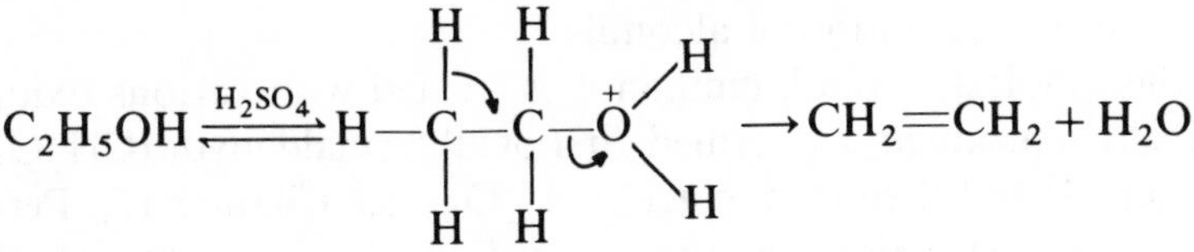

$$C_2H_5OH \underset{}{\overset{H_2SO_4}{\rightleftharpoons}} H-\overset{\overset{\displaystyle H}{|}}{\underset{\underset{\displaystyle H}{|}}{C}}-\overset{\overset{\displaystyle H}{|}}{\underset{\underset{\displaystyle H}{|}}{C}}-\overset{+}{O}\overset{\diagup H}{\diagdown_H} \rightarrow CH_2{=}CH_2 + H_2O$$

The reactions which may be involved when ethanol is treated with sulphuric acid have been commented upon in some detail because they provide an excellent example of the complications of organic chemistry. In this case ethanol and sulphuric acid may give rise to the formation of ethyl sulphate, diethyl sulphate, diethyl ether and ethylene—or a mixture of some or all of these products. Furthermore, the formation of the ether and of ethylene may involve either initial formation of ethyl sulphate or protonation of alcohol.

Such complications are commonplace in organic chemistry and present the fascinating challenge of trying to understand what set of conditions are most likely to lead to the preferential formation of a particular desired product. From the information already given it might be guessed that in the present case formation of sulphate ester, ether or alkene might be encouraged by adopting the conditions shown in the following chart:

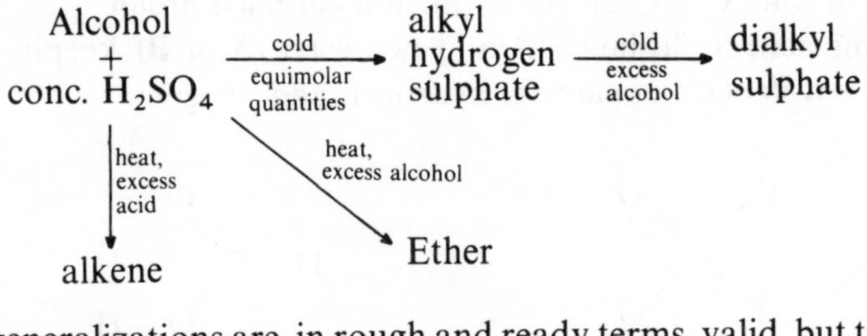

Alcohol + conc. H_2SO_4 $\xrightarrow[\substack{\text{equimolar}\\\text{quantities}}]{\text{cold}}$ alkyl hydrogen sulphate $\xrightarrow[\substack{\text{excess}\\\text{alcohol}}]{\text{cold}}$ dialkyl sulphate

(Alcohol + conc. H_2SO_4) $\xrightarrow[\substack{\text{excess}\\\text{acid}}]{\text{heat,}}$ alkene

(Alcohol + conc. H_2SO_4) $\xrightarrow[\text{excess alcohol}]{\text{heat,}}$ Ether

These generalizations are, in rough and ready terms, valid, but they are oversimplifications. Mixtures of products commonly result, and the challenge is to tune the conditions finely enough to achieve the best possible yield of the required product. This is why the recipes provided in practical texts should be followed meticulously; much care has usually been taken to supply the optimum conditions.

Oxidation of Alcohols

In organic chemistry oxidation often manifests itself either as loss of hydrogen atoms or gain of oxygen atoms. Both of these reactions are evident in the oxidation of alcohols.

If the simplest alcohol, methanol, is treated with various oxidizing agents, two products are formed: first of all formaldehyde (CH_2O) (see Chapter 14) and then formic acid (CH_2O_2) (see Chapter 15). Perhaps the commonest oxidizing agents used in the laboratory for this sequence are chromium trioxide or a dichromate, or less commonly potassium permanganante:

$$CH_3OH \xrightarrow[\substack{\text{or } K_2Cr_2O_7 \\ \text{(or } KMnO_4)}]{CrO_3} CH_2O \longrightarrow CH_2O_2$$

Formal-   Formic acid
dehyde

The first stage thus involves removal of hydrogen and the second step addition of oxygen.

There is only one possible structural formula for formaldehyde, namely

$$\begin{array}{c} H \\ \diagdown \\ \diagup \\ H \end{array} C\!=\!O$$

Note that the $\diagdown C\!=\!O$ group is called a **carbonyl group**.

Formic acid could be written in two ways (**A** or **B**), keeping to the normal covalencies of carbon, hydrogen and oxygen:

(A)   (B)

A ^{1}H-n.m.r. spectrum shows that the two hydrogen atoms are in quite different chemical environments, so structure **B** must be the correct structure.

Now consider the oxidation of ethanol. Again two products result, one by loss of hydrogen, the second by gain of oxygen:

$$C_2H_5OH \xrightarrow[(-2H)]{\text{oxidation}} C_2H_4O \xrightarrow[(+O)]{\text{oxidation}} C_2H_4O_2$$

Acetaldehyde   Acetic acid

By analogy with what happens in the case of methanol, structural formulae may be assigned to acetaldehyde and to acetic acid, as follows:

$$CH_3OH \rightarrow H_2C{=}O \rightarrow HC\overset{\displaystyle O}{\underset{\displaystyle OH}{}}$$

$$CH_3CH_2OH \rightarrow CH_3C\overset{\displaystyle O}{\underset{\displaystyle H}{}} \rightarrow CH_3C\overset{\displaystyle O}{\underset{\displaystyle OH}{}}$$

We are here assuming that since methanol and ethanol have the same functional group, the hydroxy group, they undergo similar reactions. This is a very fundamental tenet of organic chemistry.

These structures for acetaldehyde and acetic acid are indeed confirmed by n.m.r. (and i.r.) spectroscopy.

All primary alcohols, i.e. alcohols in which the hydroxy group is attached to a primary carbon atom, undergo similar reactions on oxidation.

$$RCH_2OH \xrightarrow{\text{oxidation}} RCHO \xrightarrow{\text{oxidation}} RCOOH$$

Primary alcohol   An aldehyde   A carboxylic acid

The products are called, respectively, **aldehydes** and **carboxylic acids**, and are discussed in, respectively, Chapters 14 and 15.

When a secondary alcohol is oxidized it too undergoes loss of hydrogen and gives a product known as a **ketone**; ketones are discussed in Chapter 14:

$$\begin{matrix} R' \\ \diagdown \\ C \\ \diagup\diagdown \\ R OH \end{matrix} \begin{matrix} H \\ \diagup \\ \\ \end{matrix} \xrightarrow{\text{oxidation}} \begin{matrix} R' \\ \diagdown \\ C{=}O \\ \diagup \\ R \end{matrix} \quad \text{A ketone}$$

When a primary alcohol is oxidized first to an aldehyde

$$\begin{matrix} R \\ \diagdown \\ \diagup \\ H \end{matrix} C{=}O$$

and then further to a carboxylic acid

$$\begin{matrix} R \\ \diagdown \\ \diagup \\ HO \end{matrix} C{=}O$$

the second step involves replacement of a hydrogen atom attached to a carbonyl group by a hydroxy group. In the case of a ketone there is no hydrogen atom attached to the carbonyl group. Therefore this second step is not possible and oxidation stops at the ketone.

The first step of the oxidation of primary or secondary alcohols involves removal of two hydrogen atoms, one forming part of the hydroxy group and the other attached to the carbon atom that carries the hydroxy group:

$$\underset{\diagup \quad \diagdown OH}{\overset{\diagdown \quad \diagup H}{C}} \longrightarrow \underset{\diagup}{\overset{\diagdown}{C}}{=}O$$

In the case of a tertiary alcohol there is no hydrogen atom linked to the carbon atom to which the hydroxy group is attached:

$$\begin{matrix} R' \\ | \\ R-C-OH \\ | \\ R'' \end{matrix}$$

A tertiary alcohol

Hence in this case even the first step of the above oxidation routes is not possible.

Thus tertiary alcohols are less easily oxidized than primary or secondary alcohols. However, by using stronger oxidizing agents and/or fiercer reaction conditions (e.g. heat) tertiary alcohols can be oxidized, but this involves break-up of the molecules with rupture of carbon–carbon bonds.

The progress of the oxidation of alcohols can be conveniently followed by infrared spectroscopy. Infrared spectroscopy provides less detailed information than does n.m.r. spectroscopy; its value lies especially in the fact that certain groups of atoms, or functional groups, absorb radiation at specific frequencies (for further information about infrared spectroscopy see the Appendix). Hence the presence or absence of such groups can be deduced from i.r. spectra of compounds. In the present context it is of use that hydroxy and carbonyl groups each give easily observed absorption signals. When an alcohol is oxidized the signal from its hydroxy group will be lost; this may, however, be impossible to observe since these oxidations are often carried out in aqueous solution and the water will also provide a hydroxy group signal. As oxidation proceeds a signal due to a carbonyl group appears, and by observing this it is possible to see that oxidation of the alcohol has taken place.

Generality of Reactions of Functional Groups

The properties of alcohols described above will apply to any molecule containing an R—OH group. For example, the biologically important compound cholesterol has the structural formula

Just by looking at this formula we can know that cholesterol will undergo reactions typical of a secondary alcohol (e.g. formation of esters, oxidation to a ketone, reaction with thionyl chloride) and an alkene (e.g. with bromine, potassium permanganate).

A most important feature of organic chemistry is this generality of reactions of functional groups. Functional groups are the groups attached to the hydrocarbon framework of an organic molecule, such as HO—, Cl—, H_2N— or non-alkane parts of the framework such as

$$\text{C=O} \quad \text{and} \quad \text{C=C}$$

These groups usually have similar properties whenever they occur in a molecule, so that, by looking at the structural formula of a molecule, its properties can, in good measure, be forecast.

Sometimes there may be complications if two functional groups are adjacent to one another and affect each other's behaviour. Examples will be seen later in the book. One such example is provided by the carboxylic acid or **carboxyl** group:

$$-C{\underset{\textstyle OH}{\overset{\textstyle O}{\diagup}}}$$

mentioned earlier in this chapter. The properties of its carbonyl and hydroxy groups are somewhat different from those of simple isolated carbonyl groups in aldehydes or ketones or those of hydroxy groups in alcohols. Another example, already mentioned in an earlier chapter, is that monohalogenoalkanes are different from polyhalogenoalkanes in that the latter are relatively unreactive towards nucleophiles.

Polyhydroxy Compounds

Stable compounds with two hydroxy groups attached to one carbon atom are uncommon. They usually lose a molecule of water to form a carbonyl group:

$$HO-\overset{\textstyle |}{\underset{\textstyle |}{C}}-OH \rightleftharpoons \overset{\textstyle |}{\underset{\textstyle |}{C}}=O \ + \ H_2O$$

This provides another example of properties of functional groups being affected by other adjacent functional groups, in this case a second hydroxy group. (In aqueous solutions of carbonyl compounds appreciable amounts of the dihydroxy compound may be present, in equilibrium with the carbonyl form.)

Polyhydroxy compounds with hydroxy groups attached to different carbon atoms are common and occur widely in nature. The simplest example is ethane-1, 2-diol, commonly known as **glycol**:

$$\begin{array}{c} CH_2OH \\ | \\ CH_2OH \end{array}$$

This compound is made from ethylene (see Chapter 5) and is an industrial chemical of great importance. It has also been used as a coolant; it can be used at higher temperatures than water because it has a higher boiling point (197 °C). The high boiling point is a consequence of its ability to undergo hydrogen bonding. For this reason it is also completely miscible with water. When dissolved in water it lowers the freezing point of the latter, and for this reason, and because it is not corrosive, it has considerable use as an antifreeze. It is highly toxic and therefore should on no account be confused with glycerol, which is used in various domestic products.

Glycerol occurs widely in nature combined with acids in esters, which form the fats and oils in living creatures, animal and vegetable:

$$
\begin{array}{ll}
\mathrm{CH_2OH} & \mathrm{RCOOCH_2} \\
| & | \\
\mathrm{CHOH} & \mathrm{R'COOCH} \\
| & | \\
\mathrm{CH_2OH} & \mathrm{R''COOCH_2} \\
\text{Glycerol} & \text{A triester of glycerol,} \\
\text{(or propane-1, 2, 3-triol)} & \text{as found in fats and oils}
\end{array}
$$

Because of the three hydroxy groups in the molecule, glycerol has a very high boiling point, and is very soluble in water. It is less important industrially than glycol and is obtained either as a by-product in the manufacture of soap from fats and oils, or from propene. It is used as a moistening agent because it is hygroscopic but involatile, and so retains water. It is used in this way in soaps, cosmetics, shaving creams and tobacco.

Esters made from reaction of glycerol with dicarboxylic acids have some use as plastics, while the ester it forms with nitric acid, glyceryl trinitrate, commonly called nitroglycerine, has use as an explosive.

Very many more complicated polyhydroxy compounds are important in living systems. For example, hexahydroxycyclohexanes, usually known as **inositols**, are very widely distributed in plants and animals, and play an important role in biological processes:

Inositols

Glucose

Sucrose (domestic sugar)

Sugars are polyhydroxy compounds; the commonest is glucose. The reactivity, chemical and biological, of naturally occurring compounds frequently depends crucially on the stereochemistry and shape of the molecules. A formula that shows better the shape of the glucose molecule is also given. The ring takes up the same chair shape as cyclohexane (see Chapter 2) and the attached functional groups stick outwards from the carbon atoms of the ring. (Try to make a molecular model of this compound; a three-dimensional model makes the structure more evident.)

Sucrose, which is obtained industrially from sugar cane and sugar beet, and used domestically as sugar, has a rather more complicated molecule. Sucrose is called a **disaccharide** because it consists of two simpler sugar units joined together by a linking oxygen atom. Each unit is called a **monosaccharide**. In sucrose, one unit is glucose, the other is fructose. Fructose is an isomer of glucose, and is the sweetest of all sugars; it occurs as a monosaccharide in fruits and in honey. When the two monosaccharide units are joined to form a disaccharide a molecule of water is lost. Sucrose reacts with water, in the presence either of acid or of an enzyme called sucrase, to give glucose and fructose. The relationship may be summed up by the following equation:

$$C_6H_{12}O_6 + C_6H_{12}O_6 \rightleftharpoons C_{12}H_{22}O_{11} + H_2O$$

Glucose   Fructose   Sucrose

Sugars are members of a large family of compounds called **carbohydrates**.

Other carbohydrates are made up of large numbers of monosaccharide units linked together, with loss of a molecule of water for each link. These are known as **polysaccharides**. Examples of polysaccharides are **cellulose** and **starch**, both of which are made up solely of glucose units, but differ in the way in which these units are linked together.

Cellulose is possibly the commonest naturally occurring organic compound on earth. It is the chief constituent of plant cell walls, and especially of the harder and more fibrous strengthening tissues. Cotton is almost pure cellulose; linen, jute and other plant fibres are also mostly cellulose. Wood is another important source but in this case the cellulose is mixed with **lignin**, which is not a polysaccharide. In the preparation of paper from wood this lignin has to be removed. Filter paper is almost pure cellulose.

Cellulose molecules are made up from up to 3000 glucose units and the structure consists of a long chain with a repeating pattern:

etc. — CH_2OH — HO — HO — O — CH_2OH — HO — HO — O — CH_2OH — HO — HO — O — etc.

Cellulose does not react readily with water and in consequence is not easily broken down, or **hydrolysed**, by water to produce glucose. Because of this, it is not a practicable industrial source of glucose. For the same reason it is useless as food for higher animals. Herbivorous animals can digest and use it owing to the presence of bacteria in their digestive systems that effect its hydrolysis.

Starch occurs in all green plants, the commonest sources being potatoes and cereals. It is hydrolysed to glucose by enzymes in the body and hence it has enormous importance as a foodstuff. It is fairly easily hydrolysed either by the action of dilute aqueous acid or of the enzyme diastase in malt. Industrially it is an important source of glucose, and also of ethanol obtained by the breakdown of glucose by yeast, and it is a fundamental raw material of the brewing industry. Starch differs from cellulose chemically in the way that the constituent glucose units are assembled.

All of these polyhydroxy compounds can be oxidized, and form esters. Glyceryl trinitrate has already been mentioned. Cellulose also forms nitrate esters, commonly known as guncotton, which are explosives of commercial importance. These explosives have a large civilian importance, e.g. for blasting.

Esters from acetic acid are also of great importance. Cellulose acetates are in very common use as synthetic fibres, under a number of trade names. Cellulose acetate is also used as cinema film.

The hydroxy groups of cellulose may be converted into ether groups, —OR; these cellulose ethers have wide commercial uses as dispersing agents, e.g. in food, and for controlling the viscosity of materials such as paints and pastes.

For relatively small molecules, such as glycerol, glucose or sucrose, the presence of a large number of hydroxy groups in the molecules makes them soluble in water. However, as the molecules become increasingly large, the sheer size lowers the solubility and, if large enough, they are totally insoluble. It is this insolubility that has enabled the plant world to make use of cellulose as its main structural constituent. Although cellulose is insoluble in water, because of its hydroxy groups it is hygroscopic and can absorb water. Starch is similarly insoluble in water but hygroscopic.

Questions

1. A compound contains 60% C, 13.33% H, 26.67% O. What are the ratios of the numbers of carbon, hydrogen and oxygen atoms present?

   If the molecular weight of the compound is 60, what are the possible structural formulae?

   How many ^{1}H-n.m.r. and ^{13}C-n.m.r. signals would each isomer provide?

2. 10.9 g of ethyl bromide reacted with an excess of sodium hydroxide and provided 2.3 g of ethanol.

   In an organic chemical preparation it is usual to quote the **percentage yield**, which expresses the actual yield as a percentage of the yield which is theoretically obtainable. What is the percentage yield of ethanol obtained in the present example?

   Alternative products formed in other competing reactions are called **by-products**. What by-products might be formed in the above example?

3. Why would the following reactions not take place?

(a) $CH_3CH_2CH_2CH_3 + NaOH \longrightarrow CH_3CH_2CH_2CH_2OH$
(b) $C_2H_5I + KNO_3 \longrightarrow C_2H_5ONO_2 + KI$
(c) $NaCl + C_2H_5I$ in acetone $\longrightarrow C_2H_5Cl + NaI$
(d) $(C_2H_5)_3\overset{+}{N}H \ \ ^-OH \xrightarrow{\text{heat}} C_2H_4 + (C_2H_5)_2NH$

(e) $C_2H_5OH + Na^+ \ \ ^-OH \longrightarrow C_2H_5O^- \ \ Na^+ + H_2O$

4. How could the following transformations be accomplished? Note that some cannot be carried out directly and may need more than one step.

(a) An alcohol into an ether
(b) An alternative method for (a)
(c) 2-bromobutane into butan-2-one (ethyl methyl ketone, $CH_3COCH_2CH_3$)
(d) ethanol into cyanoethane
(e) propan-2-ol into 2-aminopropane

5. How could you obtain:

(a) 2-chloropropane from propan-1-ol,
(b) 1,2-dibromo-2-methylpropane from 2-methylpropan-2-ol,
(c) 1-bromopropane from 2-iodopropane,
(d) cyclohexene from cyclohexane?

10 ETHERS

As mentioned in the previous chapter, ethers, $R—O—R'$, are isomeric with alcohols. The two simplest methods for distinguishing ethers from alcohols are (a) to record an i.r. spectrum and note whether or not there is absorption that can be attributed to the presence of a hydroxy group and (b) to test whether or not the compound gives two reactions diagnostic of the presence of a hydroxy group, namely to give off hydrogen when sodium is added and to give off hydrogen chloride on addition of phosphorus pentachloride. Ethers give neither of these reactions.

Ethers may, of course, be isomeric with one another; e.g. an ether with the molecular formula $C_4H_{10}O$ could be

$CH_3OCH_2CH_2CH_3$	or	$CH_3OCH(CH_3)_2$	or	$CH_3CH_2OCH_2CH_3$
Methyl propyl ether		Methyl isopropyl		Diethyl ether
or		ether		or
1-methoxypropane	or	2-methoxypropane		ethoxyethane

Nomenclature

Two forms of nomenclature are commonly met. For simple ethers the names of the alkyl groups attached to the oxygen atom are specified, as in ethyl methyl ether, $C_2H_5OCH_3$. This method is commonly used for simple small molecules. The more generally applicable method is to regard the ether as an alkane substituted by an **alkoxy** group, $RO—$. The larger of the two alkyl groups in the ether is treated as the alkane and the smaller as part of an alkoxy group. In this way of naming, ethyl methyl ether becomes methoxyethane. Other examples are given for the isomeric ethers discussed in the preceeding paragraph.

Preparation of Ethers

Two methods have been described in earlier chapters whereby ethers may be prepared.

One method involves reaction of a sodium alkoxide with an alkyl halide:

$$Na^+ \quad RO^- \frown R'{-}Hal \rightarrow R{-}O{-}R' + Hal^-$$

This method may be used to prepare symmetric ($R = R'$) and non-symmetric ($R \neq R'$) ethers.

Symmetric ethers only can be prepared by heating an excess of an alcohol with concentrated sulphuric acid:

$$ROH + H_2SO_4 \rightarrow R \frown OSO_3H \rightarrow R{-}\overset{+}{O}{-}R \quad \bar{O}SO_3H$$

$$R{\diagup}\overset{\displaystyle O}{\diagdown}H \qquad\qquad \overset{|}{H}$$

or

$$ROH + H_2SO_4 \rightarrow R \frown \overset{\displaystyle \overset{H}{|}}{\underset{+}{O}}{-}H \rightarrow R{-}\overset{+}{O}{-}R$$

$$R{\diagup}\overset{\displaystyle O}{\diagdown}H \qquad\qquad \overset{|}{H}$$

$$R{-}O{-}R$$

$$+ H_2O$$

Common Ethers

Diethyl ether is usually made from ethanol and, as the commonest ether, is frequently called just **ether**. It is a very good solvent for many organic compounds and is commonly used in this role. Care must be taken, however, since it is very volatile at room temperature, its boiling point being 35 °C, and the vapour is very readily ignited. Dimethyl ether is a gas at room temperature. Its boiling point is − 24 °C and it has been used as a refrigerant and also as a local anaesthetic by freezing the area to which it is applied because of heat absorbed in its evaporation. Again great care is needed to keep it away from flames.

Volatility and Solubility of Ethers

Ethers have lower boiling points than alcohols of similar molecular weight because they do not have hydroxy groups that can be hydrogen bonded to other molecules. On the other hand, ethers can accept hydrogen bonds from other molecules having hydroxy groups:

$$\underset{R}{\diagdown}O-H \cdots\cdots \underset{R'}{\diagdown}O-R''$$

For this reason ethers of relatively low molecular weight, having up to four carbon atoms, are significantly soluble in water. This solubility is not large, however, and mixtures of ether and water form two layers, water forming the bottom layer since it is denser than ether. Ethers of higher molecular weight do not dissolve to any significant extent in water and resemble alkanes more in their physical properties.

Chemical Properties

The chemical properties of alcohols that depend upon the presence of hydroxy groups are obviously not possible for ethers, e.g. they cannot be directly converted into esters or alkenes. In general ethers are not very reactive compounds. They are not easily oxidized or reduced. This unreactivity is a reason why they are used so much as solvents; they do not take part in unwanted reactions with the solutes.

Like alcohols, however, they have oxygen atoms which can be protonated, especially in strong acids, to give oxonium salts:

$$R-O-R' \xrightarrow{\ HI\ } \quad \underset{H}{\overset{R}{\diagdown}}\overset{+}{O}-R' \ \ I^-$$

Hence ethers are soluble in strong acids. Addition of water causes reformation of the ether, which comes out of solution.

These oxonium ions can undergo nucleophilic substitution by non-basic nucleophiles, i.e. by anions of strong acids. Basic nucleophiles cannot react in this way because they preferentially react just to remove the proton from the oxonium cation:

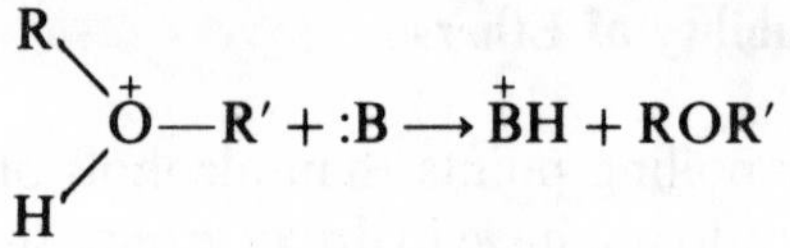

$$\underset{H}{\overset{R}{>}}\overset{+}{O}{-}R' + :B \longrightarrow \overset{+}{B}H + ROR'$$

(:B = basic nucleophile)

An example of a nucleophilic substitution reaction involving an oxonium ion derived from an ether is as follows:

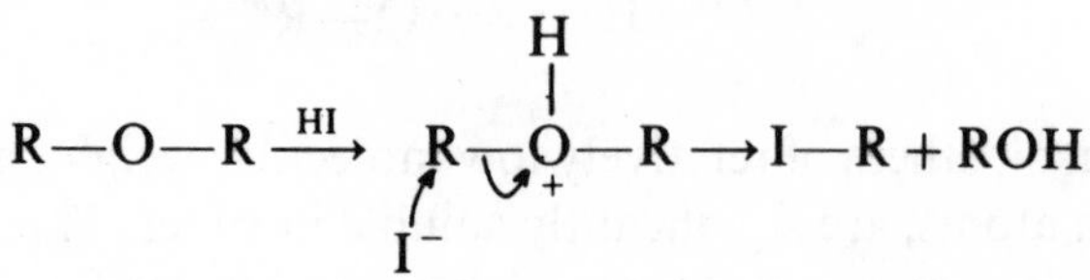

$$R{-}O{-}R \xrightarrow{HI} \underset{I^-}{R{\overset{H}{\underset{+}{{-}O}}}{-}R} \longrightarrow I{-}R + ROH$$

The alcohol, ROH, is a good leaving group. It can now react with hydriodic acid to form more alkyl iodide:

$$ROH + HI \longrightarrow RI + H_2O$$

Hence the overall reaction is

$$ROR + 2\,HI \longrightarrow 2\,RI + H_2O$$

Other strong acids react similarly, for example

$$ROR' + 2\,HBr \longrightarrow RBr + R'Br + H_2O$$

$$ROR' + 2\,H_2SO_4 \longrightarrow ROSO_3H + R'OSO_3H + H_2O$$

Some Cyclic Ethers

An oxygen atom can form part of a ring to give a cyclic ether, for example

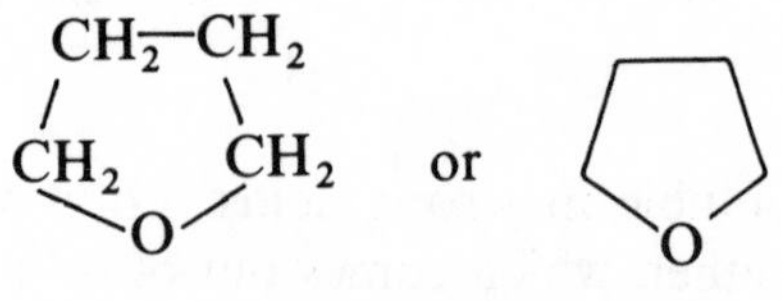

$$\underset{\underset{O}{CH_2 \qquad CH_2}}{CH_2{-}CH_2} \qquad or \qquad \text{(ring structure with O)}$$

Tetrahydrofuran

Tetrahydrofuran gets its name because it is made by catalytic reduction of furan (see Chapter 28):

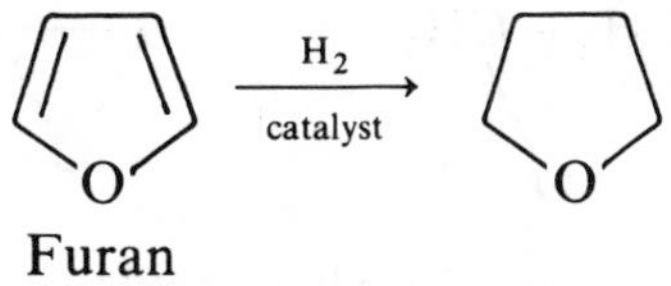

Furan

It is a very useful solvent. It is completely soluble in water because the oxygen atom and the lone pairs of electrons on this atom which participate in hydrogen bonding with water are more readily accessible than is the case with diethyl ether:

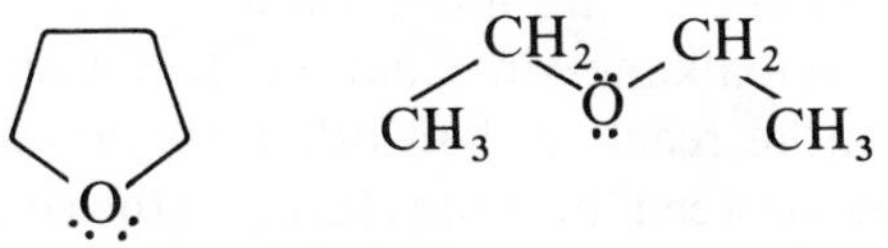

In the case of tetrahydrofuran the alkyl portion of the molecule is held back and so cannot hinder the approach of a water molecule in the way that the more mobile ethyl groups of diethyl ether can. Like non-cyclic ethers, tetrahydrofuran is protonated in strong acids, and can undergo nucleophilic substitution by non-basic nucleophiles, which leads in this case to opening of the ring:

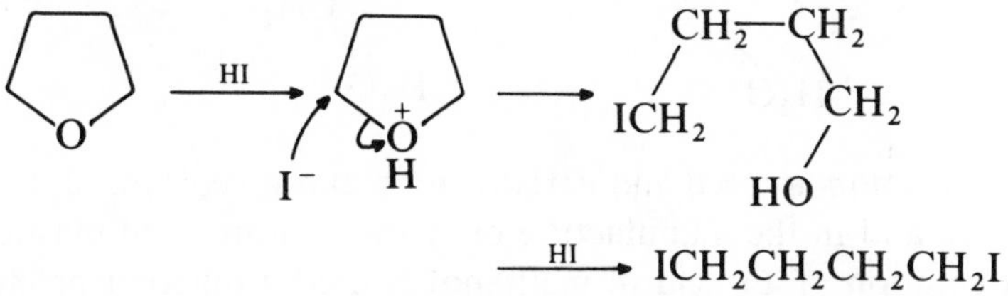

$$\xrightarrow{\text{HI}} ICH_2CH_2CH_2CH_2I$$

A ring in which one or more of the carbon atoms is replaced by another kind of atom is known as a **heterocycle.** Tetrahydrofuran is a **heterocyclic** compound. The cyclic amines mentioned in Chapter 8 are also heterocycles.

Ethylene Oxide or Oxiran

Heterocyclic rings may be of any size from three atoms upwards. The cyclic ether having a three-membered ring has the systematic name oxiran but is more commonly called ethylene oxide, a name that reflects both its source and method of manufacture:

$$\underset{\underset{\displaystyle O}{\diagdown\diagup}}{CH_2-CH_2} \qquad \text{or} \qquad \underset{O}{\triangle}$$

Ethylene oxide

As described in Chapter 5, ethylene oxide is made by oxidation of ethylene. It has a boiling point of 13.5 °C and is a compound of great industrial importance. The largest use of ethylene is in the preparation of polythene, but the next largest use is for the preparation of ethylene oxide, and millions of tons are made annually.

In discussing cycloalkanes in Chapter 2, it was mentioned that cyclopropane is more reactive than other alkanes and cycloalkanes because the angles between the bonds forming the ring are forced to be smaller than in other compounds. This introduces strain into the molecule. Similar considerations apply to ethylene oxide. The C—O bonds are broken more readily than in other ethers. For example, ethylene oxide reacts with dilute aqueous acid to form glycol:

$$\underset{CH_2}{\overset{CH_2}{\diagdown}}\!\!\!O \;\;\overset{H^+}{\underset{H_2O}{\longrightarrow}}\;\; \underset{\underset{H_2O}{\nearrow}CH_2^+}{\overset{CH_2}{\diagdown}}\!\!\!O\!-\!H \;\longrightarrow\; \underset{\underset{H_2O^+}{}CH_2}{\overset{CH_2}{\diagdown}}\!\!\!OH \;\;\overset{-H^+}{\longrightarrow}\;\; \underset{CH_2OH}{\overset{CH_2OH}{|}}$$

This reaction is used industrially to make glycol, for use as an antifreeze and in the manufacture of synthetic fibres and plastics.

If a solution of an acid in methanol is used a different product is obtained:

$$\underset{CH_2}{\overset{CH_2}{\diagdown}}\!\!\!O \;\;\overset{H^+}{\longrightarrow}\;\; \underset{\underset{H}{\diagup}CH_3\diagdown O\diagup CH_2}{\overset{CH_2}{\diagdown}}\!\!\!O\!-\!H \;\longrightarrow\; \underset{\underset{CH_3OH^+}{|}}{\overset{\overset{CH_2OH}{|}}{CH_2}} \;\;\overset{-H^+}{\longrightarrow}\;\; \underset{CH_2OCH_3}{\overset{CH_2OH}{|}}$$

This product is made for adding to jet-engine fuels, to prevent the formation of ice crystals.

When a solution of acid in glycol is used another similar reaction takes place:

$$
\begin{array}{ccc}
\underset{\mid}{CH_2} \\[-2pt]
\overset{\mid}{CH_2}
\end{array}\!\!\!\!O \xrightarrow{\ H^+\ }
\begin{array}{c}
\underset{\mid}{CH_2}\\[-2pt]
\overset{\mid}{CH_2}
\end{array}\!\!\!\!OH \longrightarrow
\begin{array}{c}
CH_2OH\\
\mid\\
CH_2\\
\mid\\
CH_2\overset{+}{OH}\\
\mid\\
CH_2OH
\end{array} \xrightarrow{\ -H^+\ }
\begin{array}{c}
CH_2OH\\
\mid\\
CH_2\\
\mid\\
O\\
\mid\\
CH_2\\
\mid\\
CH_2OH
\end{array}
$$

This compound is used as a plasticizer in the resin used to bind granules of cork to form tiles, and also as a solvent.

Ammonia attacks ethylene oxide. In this case ammonia is a strong enough nucleophile to attack the non-protonated ether:

$$
\begin{array}{c}
CH_2\\[-2pt]
\mid\!\!\!\!>O\\[-2pt]
CH_2\\
\\
:NH_3
\end{array} \longrightarrow
\begin{array}{c}
CH_2-O^-\\
\mid\\
CH_2\\
\mid\\
+\,NH_3
\end{array} \longrightarrow
\begin{array}{c}
CH_2OH\\
\mid\\
CH_2NH_2
\end{array}
$$

This compound is used to absorb carbon dioxide in the manufacture of dry-ice, solid carbon dioxide.

As a final example, hydrochloric acid reacts to give 2-chloroethanol, which is itself an important intermediate used to make a variety of other compounds:

$$
\begin{array}{c}
CH_2\\[-2pt]
\mid\!\!\!\!>O\\[-2pt]
CH_2
\end{array} +\ HCl \longrightarrow
\begin{array}{c}
CH_2OH\\
\mid\\
CH_2Cl
\end{array}
$$

These are only a few of the reactions and uses of ethylene oxide, but serve to illustrate its versatility, usefulness and industrial importance.

Other alkene oxides are made similarly and are also valuable intermediates, since they undergo similar reactions to those of ethylene oxide.

Propylene oxide is also made on a huge scale, from propene, and is a starting material in the preparation of various polymers, as is a chloro derivative of propylene oxide:

$$\overset{\text{O}}{\underset{\text{CH}_2-\text{CH}_2-\text{CH}_3}{\triangle}} \qquad \overset{\text{O}}{\underset{\text{CH}_2-\text{CH}_2-\text{CH}_2\text{Cl}}{\triangle}}$$

Propylene oxide

Crown Ethers

Crown ethers are large cyclic ethers with rings made of carbon and oxygen atoms. An example is 18-crown-6:

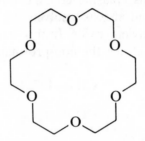

18 refers to the total number of atoms making up the ring and 6 to the number of oxygen atoms. Crown ethers can take up metal ions within the ring; the metal ions are held in place by interactions with the lone pairs of electrons on the oxygen atoms, for example,

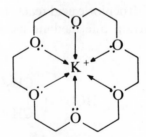

In this complex the outside ring is made up of hydrocarbon-like segments ($-\text{CH}_2\text{CH}_2-$), and this makes the complex soluble in organic solvents. In this way an ion which is normally insoluble in an organic solvent can be made soluble.

For example, if potassium permanganate is added to benzene it does not dissolve and the benzene remains colourless. When a small amount of 18-crown-6 is added to the benzene, the benzene takes on a purple colour as the crown ether takes the salt into solution. Thus this enables potassium permanganate to be used as an oxidizing agent in solution in benzene.

Obviously the relative sizes of the central 'hole' in the crown ether and of the cation are important if a complex is to be formed. The cation must be neither too big to enter the hole, nor so small that the fit is too loose. Thus 18-crown-6 readily takes up K^+, but the smaller Li^+ ion does not form a complex. Hence differently sized crown ethers may be specific for complexing with different cations, and in this way different cations in a mixture may be separated from one another.

Large-ring compounds with nitrogen atoms forming part of the ring, such as that shown below, may complex cations in an analogous fashion:

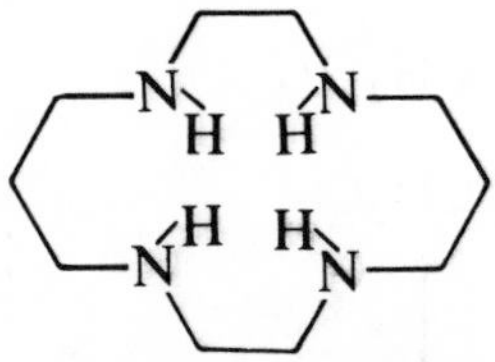

Metal complexes of compounds having large nitrogen-containing rings play a literally vital role in biological systems. Examples include vitamin B_{12}, whose formula is depicted on the last page of this text, which is effective in the treatment of pernicious anaemia, chlorophyll, the green pigment of plants, which is involved in photosynthesis in plants, and haem, which forms part of haemoglobin, which is present in the red corpuscles of blood and acts as the oxygen carrier in the body.

Questions

1. Give **four** methods, two spectroscopic and two chemical, for distinguishing between the two isomers $C_2H_5OC_2H_5$ and $CH_3CH_2CH_2CH_2OH$.
2. A compound has a molecular formula C_3H_8O. Write possible structural formulae. How would you distinguish between these isomers by ^{1}H-n.m.r. spectroscopy? How can one of the isomers be distinguished from the others by means of a chemical test?
3. Does the following reaction take place?

$$C_2H_5OC_2H_5 + H_2O \longrightarrow 2C_2H_5OH$$

Give reasons for your answer.

4. A compound X, $C_5H_{12}O$, may be dried over sodium. In its ^{13}C-n.m.r. spectrum there are three signals, and in its ^{1}H-n.m.r. spectrum there are two signals in the ratio 3:1.

   It reacts with hydrogen iodide to give two products (Y and Z), each of which has only one signal in its ^{1}H-n.m.r. spectrum. One of these products (Y) has only one signal in its ^{13}C-n.m.r. spectrum, the other (Z) has two signals in its ^{13}C-n.m.r. spectrum. Assign structural formulae to X, Y and Z. Explain the information provided. Show by means of equations how X could be prepared from an alcohol and/or an alkyl halide.

5. What reagents could be used to carry out each of the following reactions?

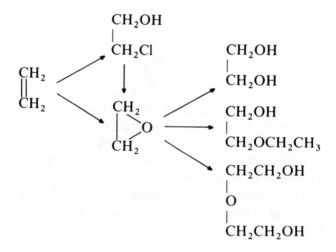

11 A COMPARISON OF COMPOUNDS CONTAINING HALOGEN, OXYGEN OR NITROGEN FUNCTIONAL GROUPS

As a tailpiece to chapters dealing with organic compounds having as functional groups halogen atoms (alkyl halides), oxygen (alcohols, ethers) and nitrogen (amines), it seems worthwhile to sum up properties of these compounds by comparing them briefly with one another.

Because of the electronegativities of the heteroatoms (i.e. atoms other than carbon or hydrogen) present in all of these compounds, they all have polarized bonds between a carbon atom and the heteroatom.

The effectiveness of the different substituent groups as leaving groups in nucleophilic substitution (S_N) reactions can be summarized as follows:

$R-Cl \rightarrow R^+ + Cl^-$ Cl^- is a good leaving group.

$R-OH \rightarrow R^+ + {}^-OH$ ⎱ These reactions do not normally take place.
$R-OR \rightarrow R^+ + {}^-OH$ ⎰ ^-OH and ^-OR only leave in protonated form as H_2O or ROH, that is

$$ROH + H^+ \rightleftharpoons R\overset{+}{O}H_2 \rightarrow R^+ + H_2O$$

$$ROR + H^+ \rightleftharpoons R_2\overset{+}{O}H \rightarrow R^+ + HOR$$

$R-NH_2$ $^-NH_2$ does not act as a leaving group.

Thus the reactivities of these compounds with nucleophiles can be correlated with the leaving group ability of the functional group, viz.

R—Cl   React readily
R—OH, R—OR   React only when protonated
R—NH$_2$ Does not undergo nucleophilic substitution

These differences in reactivity in S$_N$ reactions may be compared with the properties of compounds in which the alkyl group R is replaced by a hydrogen atom:

H—Cl $\rightleftharpoons$ H$^+$ + Cl$^-$ Cl$^-$ is a good leaving group; HCl is a fairly strong acid.

H—OH $\rightleftharpoons$ H$^+$ + $^-$OH   HO$^-$ and RO$^-$ are poor leaving groups;

(or H—OR $\rightleftharpoons$ H$^+$ + $^-$OR)   HOH and ROH are weak acids.

H—NH$_2$ H$^+$ is only removed by very strong bases.

Another characteristic of nitrogen, oxygen and halogen atoms is that they all have lone pairs of electrons not involved in chemical bonding. These lone pairs of electrons may take part in nucleophilic attack on other molecules such as alkyl halides, for example

Since it might be expected that ability to form covalent bonds should be the converse of effectiveness as a leaving group, it should also be expected that ability to form covalent bonds will increase in the order Cl < OR$'$ < NR$'_2$ (R$'$ = H or R). Hence the ability to react as nucleophiles should increase in the order R$'$Cl < R$'$OR$'$ < NR$'_3$ (R$'$ = H or R). This is so. Ammonia and amines are stronger nucleophiles than are water, alcohols or ethers; covalent hydrogen halides or alkyl halides are not nucleophiles, i.e. as nucleophiles,

$$\text{Amines} > \frac{\text{water}}{\text{alcohols}} > > > \text{alkyl halides}$$

The properties of these classes of compounds, represented generally as C—X, may be summed up by the comparative importance of the leaving group ability of X and the reactivities of the lone pairs of electrons on X:

1. For halides, the most important contributing factor is the inductive effect $C \rightarrow X$.
2. For amines, the most important contributing factor is the availability of the lone pair of electrons, CX:.
3. For alcohols and ethers, the position is intermediate between those of halides and amines.

In diagrammatic form:

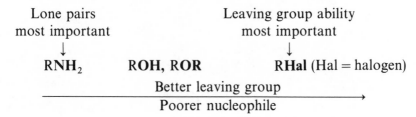

Comparative properties of $C \rightarrow X$:

Lone pairs most important		Leaving group ability most important
↓		↓
RNH_2	**ROH, ROR**	**RHal** (Hal = halogen)

Better leaving group →
Poorer nucleophile

Question

1. Starting from compounds of formula RCH_2CH_2X, wherein X is any group of your choice, list, with mechanisms, three general methods for the preparation of alkenes $RCH = CH_2$.

12 SULPHUR ANALOGUES OF ALCOHOLS AND ETHERS: THIOLS AND THIOETHERS

Since sulphur comes just below oxygen in the periodic table, sulphur analogues of alcohols and ethers should resemble the oxygen-containing compounds.

The prefix **thio-** signifies sulphur. A **thiol** is a sulphur analogue of an alcohol, R—SH (alcohols are denoted by the suffix -ol) and R—S—R' is a **thioether**. Not infrequently thiols and thioethers are called by the older names, respectively, **mercaptan** and **sulphide.**

Thiols

A typical thiol is CH_3CH_2SH, ethanethiol or ethyl mercaptan. One of the most obvious characteristics of thiols is their unpleasant sweet smell, which is very powerful. The human nose can detect ethanethiol in a concentration of only one part in 10^9. Butane-1-thiol, $CH_3CH_2CH_2CH_2SH$, is a contributant to the odour of skunks.

Thiols are made from alkyl halides:

$$Na^+ HS^- + RCl \longrightarrow HSR + Cl^- (+ Na^+)$$

Thiols are more acidic than alcohols and form salts when treated with hydroxides:

$$R—SH + HO^- Na^+ \longrightarrow R—S^- \quad Na + H_2O$$

They are readily oxidized to disulphides, which are, in turn, readily reduced again to thiols:

$$RSH \xrightarrow{\text{oxidation}} R—S—S—R \xrightarrow{\text{reduction}} RSH$$

A disulphide

This reaction is extremely important in the chemistry of proteins and in natural biological processes. Some proteins contain thiol groups. Oxidation of such proteins links two protein chains together:

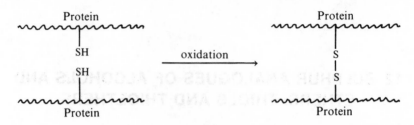

It has been suggested that this reaction may be involved in the mechanism of memory in the brain. On a more mundane level this process is used in hair-perming. In the treatment of the hair, disulphide links between proteins are first broken and then new disulphide links are formed in other positions of the protein chains. The breaking of disulphide links is also involved in the kneading of dough.

Alkyl Sulphides

Alkyl sulphides are made by the reaction of sodium salts of thiols with alkyl halides:

$$RS^- Na^+ + R'Hal \longrightarrow RSR'$$

An infamous example is bis(2-chloroethyl)sulphide, $ClCH_2CH_2S\text{-}CH_2CH_2Cl$, which is mustard gas.

Sulphides are oxidized to form **sulphoxides**, for example

$$RSR + H_2O_2 \longrightarrow R\overset{\displaystyle O}{\overset{\displaystyle \|}{-}}S-R$$

A sulphoxide

Dimethyl sulphodixe, $(CH_3)_2SO$, is a very useful solvent and can dissolve many inorganic salts as well as organic compounds. It is made industrially by oxidation, by air, of dimethyl sulphide.

Sulphides react with alkyl halides to give **sulphonium salts**:

$$R_2S \frown R \overset{\frown}{-} I \longrightarrow R_3S^+ \ I^-$$

A sulphonium salt

13 ORGANOMETALLIC COMPOUNDS

In the foregoing chapters the organic compounds that have been considered have had only a limited range of elements attached to a carbon atom, namely hydrogen, the halogens, nitrogen, oxygen and sulphur. Can carbon form bonds to other elements as well? In fact most elements can form bonds to carbon, but the chemistry of many of the compounds involved is complex. Also many **organometallic compounds**, in which a metal atom is directly attached to a carbon atom, are extremely sensitive to air and moisture, which makes many of them difficult to handle. As just one example, the aluminium compound $Al_2(CH_3)_6$ inflames on exposure to air. Other organometallic compounds are, however, more stable. Such a compound, which is commonly encountered, is tetramethylsilane, $(CH_3)_4Si$; this compound is used as a standard marker when n.m.r. spectra are recorded. (It is commonly given the nickname TMS and referred to in this way.) The signal arising from its hydrogen atoms has been assigned a chemical shift of zero ($\delta = 0$), and the positions (chemical shifts) of other signals are measured in terms of their separation from this TMS signal.

The chemistry of organic compounds containing halogens, oxygen or nitrogen is dominated by the polarity of the bond linking a carbon atom to the substituent group. This in turn results in this carbon atom bearing a partial positive charge, which is expressed as

$$C \longrightarrow Cl \quad \text{or} \quad \overset{\delta+}{C} - \overset{\delta-}{Cl}$$

For this reason the carbon atom reacts readily with nucleophiles, and this is a characteristic of the chemical behaviour of these compounds.

It seems reasonable that other bonds to carbon might result in a partial negative charge on the carbon atom:

$$C \longleftarrow X \quad \text{or} \quad \overset{\delta-}{C} \text{—} \overset{\delta+}{X}$$

For this situation to arise atom X would have to be less electronegative than carbon, i.e. to have a lower affinity for electrons. In general terms, elements that appear on the right-hand side of the periodic table have a high affinity for electrons and readily form anions (e.g. the halogens), whereas those appearing on the left-hand side of the table readily lose electrons to form cations (e.g. the alkali metals).

It follows, therefore, that a compound containing a bond such as C—Li should provide a carbon atom bearing a partial negative charge, which should, in consequence, react readily with **electrophiles**. This is the case, and **organolithium** and **organomagnesium** compounds in particular are widely used for this purpose. Since these organolithium and organomagnesium compounds are themselves readily made from alkyl halides, it means that the latter compounds are even more valuable in organic synthesis; they themselves react with nucleophiles, and they are easily converted into compounds that react with electrophiles.

The chemistry of organometallic compounds is extensive, complicated and important. Despite its importance, and because of its complexity, it will be dealt with only briefly in this text.

Preparation of Organometallic Compounds

The two commonest methods for the preparation of organometallic compounds start from alkyl halides or from other organometallic compounds.

Examples of the first method are the preparation of magnesium and lithium compounds:

$$RBr + Mg \longrightarrow RMgBr$$
$$RI + 2\,Li \longrightarrow RLi + Li^+\,I^-$$

When the second method is used the commonest starting materials are organolithium or organomagnesium compounds. Some examples are as follows:

$$CH_3Li + CuCl \longrightarrow CH_3Cu + Li^+\,Cl^-$$
$$4\,CH_3MgCl + SiCl_4 \longrightarrow (CH_3)_4Si + 4\,MgCl_2$$
$$4\,C_2H_5MgCl + 2\,PbCl_2 \longrightarrow (C_2H_5)_4Pb + Pb + 4\,MgCl_2$$

Copper compounds find use in organic synthesis, while the latter two products, tetramethylsilane and tetraethyl lead, are of commercial importance. As mentioned above, tetramethylsilane is routinely used to provide the marker signal in n.m.r. spectroscopy.

Tetraethyl lead has had wide use as an additive to petrol, as an 'anti-knock'; it has the effect of lowering engine knock by preventing too fast an explosion of the air fuel mixture. Its use is now being phased out, on environmental grounds, because of lead pollution arising from exhaust fumes of cars running on petrol which contains it. Its complete disappearance will, however, take time, since many car engines have been designed in the past on the assumption that it would be present in the petrol they consumed, and in its absence they do not operate satisfactorily. Industrially it is made by reaction of chloroethane with a lead–sodium alloy:

$$4\,C_2H_5Cl + PbNa \longrightarrow Pb(C_2H_5)_4 + 4\,Na^+\,Cl^-$$

Apart from tetramethylsilane, other organosilicon compounds are of industrial importance, in particular polymeric compounds with the structure

$$\left(\begin{array}{c}R \\ | \\ -Si-O-Si-O-Si-O- \\ | \\ R\end{array}\right)_n$$

(R = alkyl group)

These compounds are known as **silicones**. In the laboratory they are prepared from magnesium compounds:

$$2\,RMgX + SiCl_4 \longrightarrow SiR_2Cl_2 \xrightarrow{\;H_2O\;} silicone$$

(X = halogen)

Industrially they are made by heating an alkyl halide with a copper–silicon alloy:

$$RCl + SiCu\ alloy \xrightarrow{\;heat\;} SiR_2Cl_2 \xrightarrow{\;H_2O\;} silicone$$

By varying the alkyl groups, the properties of the silicone can be modified. In general silicones are non-volatile and are resistant to high temperatures. They are also fairly resistant to acids and bases. Because of these properties they are compounds of considerable industrial importance, being used in water-proofing materials, for making water-proof films, in polishes and as high-temperature lubricants.

Organomagnesium Compounds. Grignard Reagents

Compounds occasionally acquire the name of the man who discovered them or exploited their use. For this reason organomagnesium compounds are commonly called **Grignard reagents**, after the French chemist who demonstrated and exploited their versatility as chemical reagents, for which he was awarded a Nobel prize in 1912.

The formula given above, RMgBr, is not the complete representation of the total structure of a Grignard reagent; the full structure is more complex and also involves the solvent in which they are made, which is an ether. Thus the formula RMgBr is in the nature of a cartoon, which, rather than giving the exact picture of the structure of the molecule, serves to show up the salient points in its character. A Grignard reagent has polarities as shown:

$$\overset{\delta-}{R} - \overset{\delta+}{Mg} - \overset{\delta-}{Br}$$

the important feature being that the alkyl group bears a partial negative charge.

Grignard reagents are very sensitive to air and to moisture, and in their preparation and handling it is essential that **everything** involved— reagents, solvent, apparatus and atmosphere—**must** be absolutely dry. They are usually prepared from alkyl bromides rather than alkyl chlorides or iodides. The reaction of alkyl chlorides with magnesium is much slower and is often too sluggish for convenience. Alkyl iodides are much more expensive than alkyl bromides. Methyl iodide is usually used for making a methyl magnesium derivative since it is a liquid at room temperature whereas methyl chloride and methyl bromide are gases. The commonest method of preparation involves stirring a suspension of magnesium turnings in a solution of an alkyl halide in dried ether; an atmosphere of dried nitrogen or argon is maintained by bubbling the gas into the reaction flask:

$$Mg + RHal \xrightarrow[\text{dry } N_2 \text{ or Ar}]{\text{dry } Et_2O} RMgHal$$

(Hal = halogen)

Reactions of Grignard Reagents

Grignard reagents take part in a variety of reactions with electrophiles, of which perhaps the commonest are those with compounds having carbonyl (C O) groups.

The preparation of aldehydes (RCHO) and ketones (RCOR′) by the oxidation of alcohols was described in Chapter 9. Both of these classes of compounds have a carbonyl group in their structure. Because of the greater electronegativity of oxygen compared to carbon the carbonyl group is polarized, with a partial positive charge on the carbon atom and a partial negative charge on the oxygen atom:

$$\begin{array}{c} R \\ \searrow ^{\delta+} ^{\delta-} \\ C{=}O \\ \nearrow \\ R' \end{array}$$

Because of this charge distribution a carbonyl group reacts readily with a Grignard reagent. The alkyl group of the latter is transferred to the carbon of the carbonyl group. Thus, for a ketone,

$$\overset{\delta-}{R}\!-\!\overset{\delta+}{Mg}\!-\!Br$$
$$+$$
$$\begin{array}{c} R' \\ \searrow ^{\delta+}^{\delta-} \\ C{=}O \\ \nearrow \\ R'' \end{array}$$

$$\longrightarrow \quad R'\!-\!\underset{\underset{R''}{|}}{\overset{\overset{R}{|}}{C}}\!-\!OMgBr \xrightarrow[H_2O]{H^+} R'\!-\!\underset{\underset{R''}{|}}{\overset{\overset{R}{|}}{C}}\!-\!OH$$

$$+ \; MgBrOH$$

The detail of this reaction is complex and only the overall result is given. When the first-formed product is treated with aqueous acid it is converted into a **tertiary alcohol**.

Similarly, from an aldehyde (except formaldehyde, H_2CO), a **secondary alcohol** is formed:

$$\begin{array}{c} R' \\ \searrow \\ C{=}O \\ \nearrow \\ H \end{array} + \; RMgBr \; \longrightarrow \; R'\!-\!\underset{\underset{H}{|}}{\overset{\overset{R}{|}}{C}}\!-\!OMgBr \xrightarrow[H_2O]{H^+} R'\!-\!\underset{\underset{H}{|}}{\overset{\overset{R}{|}}{C}}\!-\!OH$$

Formaldehyde provides a **primary alcohol:**

$$\begin{array}{c} H \\ \searrow \\ C{=}O \\ \nearrow \\ H \end{array} + \; RMgBr \; \longrightarrow \; H\!-\!\underset{\underset{H}{|}}{\overset{\overset{R}{|}}{C}}\!-\!OMgBr \xrightarrow[H_2O]{H^+} H\!-\!\underset{\underset{H}{|}}{\overset{\overset{R}{|}}{C}}\!-\!OH$$

Grignard reagents react with carbonyl groups because the former are nucleophiles and the latter are electrophiles. They do not react with carbon–carbon double bonds because the latter are themselves nucleophilic.

Another example of a reaction of a Grignard reagent with a carbonyl group is their reaction with carbon dioxide:

$$\overset{\delta-}{O}=\overset{\delta+}{C}=\overset{\delta-}{O} + RMgBr \longrightarrow O=C\overset{OMgBr}{\underset{R}{<}} \xrightarrow[H_2O]{H^+} O=C\overset{OH}{\underset{R}{<}}$$

This provides a simple way of converting an alkyl halide into a **carboxylic acid**:

$$RBr \xrightarrow[\substack{dry\ Et_2O,\\ dry\ N_2}]{Mg} RMgBr \xrightarrow[(2)\ H^+,\ H_2O]{(1)\ CO_2} RCOOH$$

Carboxylic acids contain a carbonyl group, but they react with Grignard reagents in a quite different way:

$$R'COOH + RMgBr \longrightarrow RH + RCOO^-\ Mg^{2+}\ Br^-$$

The hydrogen atom which forms part of the carboxylic acid group is quite strongly acidic and is therefore a strong electrophile. In consequence, this hydrogen atom, rather than the carbonyl group of the carboxylic acid, reacts with the alkyl group of the Grignard reagent.

The reaction of water with a Grignard reagent takes a similar course:

$$RMgBr + HOH \longrightarrow RH + MgBrOH$$

Any compound that has an acidic hydrogen atom, even as weak an acid as water, reacts in this way with a Grignard reagent to form an **alkane**.

This reactivity has its uses. For example, it provides a route for converting an alkyl halide into an alkane:

$$RHal \xrightarrow[\substack{dry\ Et_2O\\ dry\ N_2}]{Mg} RMg\,Hal \xrightarrow{H_2O} RH$$

It is sometimes useful as a method to introduce a deuterium atom into a molecule. This can be achieved by using heavy water (deuterium oxide) instead of ordinary water to react with a Grignard reagent:

$$RHal \xrightarrow[\substack{dry\ Et_2O,\\ dry\ N_2}]{Mg} RMg\,Hal \xrightarrow{D_2O} RD$$

Grignard reagents are very strong bases. In consequence any solvent that is even very weakly acidic must be avoided in their presence.

Although in general ethers are safe solvents for Grignard reagents and do not react with them, one exception is the cyclic ether ethylene oxide. As has been seen in Chapter 10, this ether is unusually reactive. The reaction proceeds as follows:

$$\underset{\underset{R-MgBr}{\underset{\delta- \quad \delta+}{\uparrow}}}{\overset{O}{\underset{CH_2-CH_2}{\bigtriangleup}}} \longrightarrow \overset{OMgBr}{\underset{CH_2-CH_2R}{\diagup}} \xrightarrow[H_2O]{H^+} HOCH_2CH_2R$$

This reaction is useful as a means of lengthening a carbon chain by two atoms, as summarized by the sequence

$$RBr \xrightarrow[\substack{dry\ Et_2O, \\ dry\ N_2}]{Mg} RMgBr \xrightarrow[(2)\ H^+,H_2O]{(1)\ \overset{O}{\overset{\bigtriangleup}{CH_2-CH_2}}} RCH_2CH_2OH$$

Organolithium Compounds and Other Alkali–Metal Compounds

The preparation of organolithium compounds, RLi, has been mentioned earlier in this chapter. As in the case of Grignard reagents, their structures are not as simple as the formulae routinely used for them imply. For example, the correct molecular formula for methyl lithium is $(CH_3Li)_4$, but it is normally expressed in its cartoon form as CH_3Li.

Alkyl lithium molecules have very polar, C—Li bonds. Hence they are very reactive towards water and oxygen, which must both be rigorously excluded when an alkyl lithium is prepared and used.

Alkyl lithium derivatives resemble Grignard reagents very closely and undergo very similar reactions, for example

$$RLi + \underset{R''}{\overset{R'}{\diagdown}}C=O \rightarrow \underset{R''}{\overset{R}{\underset{|}{\overset{|}{R'-C-OLi}}}} \xrightarrow[H_2O]{H^+} \underset{R''}{\overset{R}{\underset{|}{\overset{|}{R'-C-OH}}}}$$

$$RLi + CO_2 \rightarrow RCOO^- Li^+ \xrightarrow[H_2O]{H^+} RCOOH$$
$$RLi + H_2O \rightarrow RH + LiOH$$

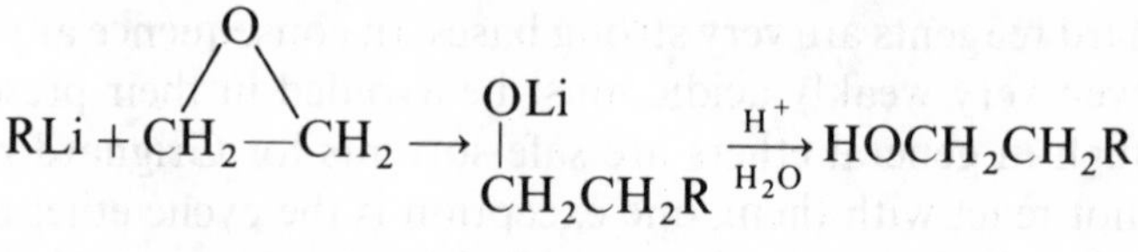

$$RLi + CH_2\!-\!CH_2 \xrightarrow{} \underset{CH_2CH_2R}{\overset{OLi}{|}} \xrightarrow[H_2O]{H^+} HOCH_2CH_2R$$

Indeed Grignard reagents and the lithium alkyls are equally applicable as reagents in the above reactions.

Alkyl sodium and alkyl potassium derivatives are even more polar than alkyl lithium derivatives, and are in fact salt-like. They are extremely reactive and sensitive to moisture and to air, so much so that they can be extremely difficult to handle.

Other Organometallic Compounds

Almost all metals can form organometallic compounds—some stable, others not; some highly reactive, others not. In recent years a large variety of these compounds has been prepared, and more and more use of them is being made. Many organometallic compounds besides those of magnesium and lithium are finding increasing use as synthetic reagents. Others are being used in industrial chemistry as catalysts.

Nature has millions of years more experience in carrying out chemical reactions than has man, and makes ample use of organometallic compounds. Some simple, but literally vital, examples (see also Chapter 10) are chlorophyll, the green pigment of plants, which contains a magnesium atom in its structure, haem, the red-coloured constituent of blood, which contains an iron atom, and vitamin B_{12}, which contains a cobalt atom.

Questions

1. Write equations, giving the mechanism if possible, to illustrate the conversion of a ketone into an alcohol by means of a Grignard reagent.
2. What reagents would you use to carry out the following reactions?

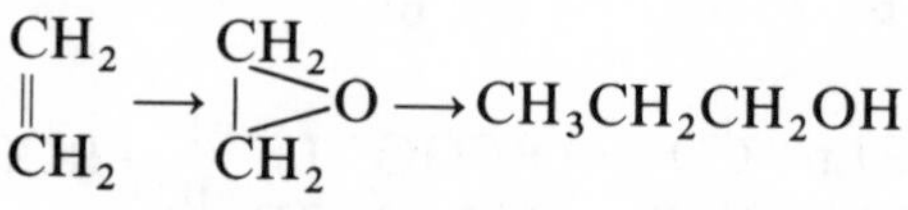

$$\underset{CH_2}{\overset{CH_2}{\|}} \rightarrow \underset{CH_2}{\overset{CH_2}{\diagup\!\!>\!\!O}} \rightarrow CH_3CH_2CH_2OH$$

3. How many signals would there be in the ^{1}H-n.m.r. and ^{13}C-n.m.r. spectra of the products of the following reactions?

(a) $CH_3CHO \xrightarrow[\text{(ii) } H^+,H_2O]{\text{(i) } C_2H_5MgBr}$

(b) $CH_3CH_2CH_2Br \xrightarrow[\text{(ii) } H_2O]{\text{(i) } Mg/(C_2H_5)_2O/N_2}$

(c) $CH_3CH_2CH_2Br \xrightarrow[\substack{\text{(ii) } CO_2 \\ \text{(iii) } H^+,H_2O}]{\text{(i) } Li}$

14 ALDEHYDES AND KETONES

When primary or secondary alcohols are oxidized the products are, respectively, **aldehydes** and **ketones**:

$$\begin{array}{c} R \\ \diagdown \\ \diagup \\ H \end{array} C{=}O \qquad\qquad \begin{array}{c} R \\ \diagdown \\ \diagup \\ R' \end{array} C{=}O$$

An aldehyde   A ketone

The functional group in both aldehydes and ketones is the **carbonyl** group, C=O. Aldehydes and ketones differ from one another in that ketones have two alkyl groups attached to the carbonyl group, whereas aldehydes have only one alkyl group attached, together with a hydrogen atom. It is possible to have a cyclic ketone in which the two groups joined to the carbonyl group are themselves joined together in a ring, for example

$$\begin{array}{c} CH_2{-}CH_2 \\ \diagup \qquad\qquad \diagdown \\ CH_2 \qquad\qquad C{=}O \\ \diagdown \qquad\qquad \diagup \\ CH_2{-}CH_2 \end{array} \qquad \text{or} \qquad \bighexagon{=}O$$

Nomenclature

The names of aldehydes are obtained by replacing the '-e' ending of the related alkane (or alkene) from which they are derived by the ending

'-al'. Examples are

$CH_3CH_2CH_2CHO$ (From $CH_3CH_2CH_2CH_3$
Butan*al*   butan*e*)
$CH_3CH{=}CHCHO$ (From $CH_3CH{=}CHCH_3$
But-2-en*al*   but-2-en*e*)

Note that the aldehyde group does not need numbering; its structure demands that it is at the end of a chain. The remainder of the chain is always numbered starting from the aldehyde group. For example,

$$\overset{\displaystyle CH_3}{\underset{}{\vert}}$$
$$\underset{5\quad\ 4\quad\ 3\quad\ 2\quad\ 1}{H_3C{-}C{=}CH{-}CH_2{-}CHO}$$

4-Methylpent-3-enal

The lower members of this series, $HCHO$ and CH_3CHO, and to a lesser extent CH_3CH_2CHO, are known by their trivial but official names formaldehyde, acetaldehyde and propionaldehyde, rather than by the systematic names methanal, ethanal and propanal. These trivial names related the aldehydes to the carboxylic acids they give when they are oxidized: formic acid, acetic acid and propionic acid.

A ketone is signified by the ending '-one'. Examples are

$CH_3CH_2COCH_2CH_3$ Pentan-3-one
$CH_3CH_2CH_2COCH_3$ Pentan-2-one

Cyclohexanone

The simplest member of the series, CH_3COCH_3, is always known by its trivial, official, name of **acetone**.

Note that pentan-2-one and pentan-3-one are isomers; a further isomer is pentanal. It should be evident that it would be simple to distinguish between these isomers by means of n.m.r. spectroscopy.

Aldehydes are particularly recognizable from their ^{1}H-n.m.r. spectra because the hydrogen atom attached to the carbonyl group, RC*H*O, provides a signal in a region of the spectrum (δ 9.3–10.0) where very few other kinds of hydrogen atoms provide signals. This is of great value in distinguishing an aldehyde from an isomeric ketone. Thus propionaldehyde, CH_3CH_2CHO, shows three signals in its ^{1}H-n.m.r. spectrum corresponding to one, two and three hydrogen atoms respectively, the first of these appearing at the appropriate position for an aldehydic hydrogen atom, whereas isomeric acetone gives only one signal in its spectrum (see next page for the spectrum).

Similarly, in their ^{13}C-n.m.r. spectra, propionaldehyde and acetone produce, respectively, three and two signals, corresponding to the numbers of different kinds of carbon atoms in the molecules (see p. 148).

The presence of the carbonyl group is usually easily detectable by an infrared spectrum. Simple aldehydes provide a signal at 1740–1720 cm^{-1} and simple ketones provide a signal at 1725–1705 cm^{-1} (see p. 147).

Polarity of the Carbonyl Group

Because of the greater electronegativity of oxygen compared to carbon, carbonyl groups are polarized with a partial positive charge on the carbon atom and a partial negative charge on the oxygen atom:

$$\overset{\delta+}{C}=\overset{\delta-}{O}$$

The reactive site is commonly the carbon atom; because of its partial positive charge it is attacked by nucleophiles. In contrast to the alkyl halides or alcohols, a nucleophile may attach itself to the carbon atom of the carbonyl group without displacing the oxygen atom:

(Nu$^-$ = nucleophile)

Instead two electrons of the double bond are displaced onto the oxygen atom which thereby acquires a negative charge but remains attached to the carbon atom by a single bond.

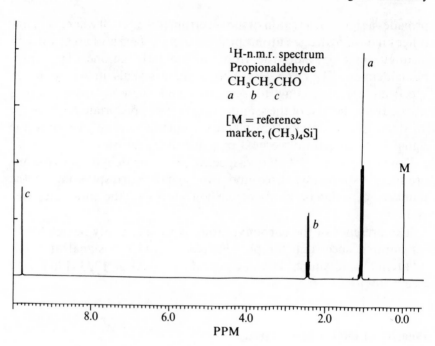

¹H-n.m.r. spectrum
Propionaldehyde
CH_3CH_2CHO
a b c

[M = reference
marker, $(CH_3)_4Si$]

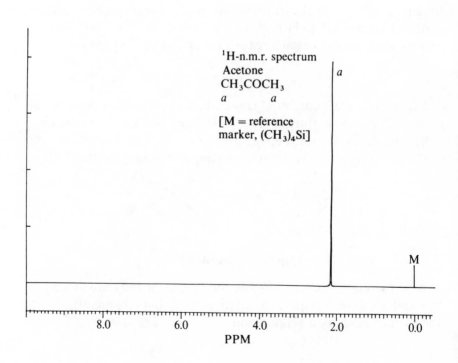

¹H-n.m.r. spectrum
Acetone
CH_3COCH_3
a a

[M = reference
marker, $(CH_3)_4Si$]

Infrared Spectrum   Frequency (cm^{-1})

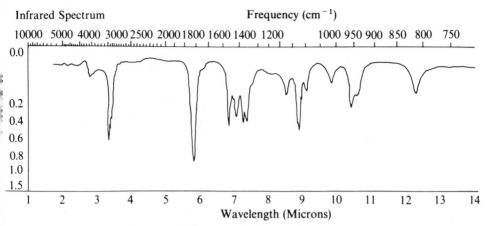

Infra-red spectrum of pentan-3-one.
Note conspicuous signal for carbonyl group at 1715 cm^{-1}

A consequence of the polarity of the carbonyl group is that it can form hydrogen bonds with water:

$$\begin{matrix} \diagdown \\ \diagup \end{matrix} C{=}O \cdots\cdots H{-}O{-}H$$

Because of this, lower members of the series are soluble in water; acetaldehyde and acetone are completely so. As the size of the alkyl groups increases so this gradually lowers the solubility in water. Because of the fact that ketones have a dual character, partly polar, at the carbonyl group, and partly non-polar, in the alkyl groups, they are very versatile and good solvents.

Formaldehyde is a gas at room temperature (b.p. $-21\,°C$), but is extremely soluble in water and is readily available in aqueous solution. Acetaldehyde also has a low boiling point ($+21\,°C$). Acetone, the simplest ketone, boils at $56\,°C$. As the number of carbon atoms in aldehydes or ketones increases, the boiling points become higher.

Uses

Major uses of ketones are as solvents. Both aldehydes and ketones are valuable chemical intermediates in industrial processes. Formaldehyde

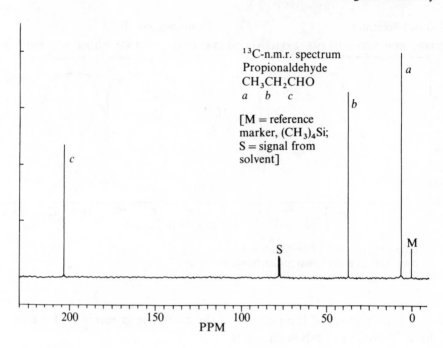

^{13}C-n.m.r. spectrum
Propionaldehyde
CH_3CH_2CHO
a   b   c

[M = reference
marker, $(CH_3)_4Si$;
S = signal from
solvent]

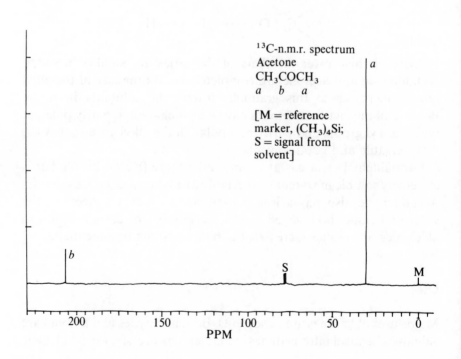

^{13}C-n.m.r. spectrum
Acetone
CH_3COCH_3
a   b   a

[M = reference
marker, $(CH_3)_4Si$;
S = signal from
solvent]

is used extensively in the manufacture of plastics. In aqueous solution, as formalin, it is used as a biological preservative for specimens; this use derives from the fact that it is a strong disinfectant and antiseptic.

Aldehydes and ketones occur profusely in nature, and are often contributants to the odour and taste of plants, etc. Some examples, chosen at random from the huge range available, are camphor, vanillin (from vanilla), menthone (mint), salicylaldehyde (meadowsweet), muscone (musk deer), citral (oil of lemon grass and many other fragrant oils of plants; it is also a honeybee pheromone). Vanillin and salicylaldehyde are derivatives of benzaldehyde (see Chapter 27). Muscone contains a large, fifteen-membered ring.

Menthone

Citral   Camphor

Both citral and menthone have ten carbon atoms and contain a substituent methyl group and a $>C(CH_3)_2$ group of atoms. They are examples, one (menthone) cyclic, the other (citral) acyclic, of a very large family of naturally occurring compounds called **terpenes**. Camphor is an example of a **bicyclic** terpene, having two rings in its structure, the $>C(CH_3)_2$ group of atoms forming a bridge across a six-membered ring. Terpenes include hydrocarbons and alcohols as well as aldehydes and ketones.

Preparation

In the laboratory aldehydes and ketones are usually made by the oxidation of alcohols, as discussed in Chapter 9.

Industrially they are made by the oxidation of alcohols or of alkenes. An example of the former process is the preparation of formaldehyde by the oxidation of methanol, using silver as a catalyst:

$$CH_3OH \xrightarrow[600-650°C]{O_2,Ag} HCHO$$

It is made in very large quantities, especially for making plastics, particularly phenol-formaldehyde resins such as **Bakelite**, which was the first synthetic polymer to be made commercially.

Oxidation of alkenes makes use of palladium chloride as a catalyst. In the process this forms an organopalladium compound as an intermediate. An example of this process is the conversion of ethylene into acetaldehyde:

$$CH_2{=}CH_2 \xrightarrow[H_2O]{O_2,PdCl_2} CH_3CHO$$

[Some copper(II) chloride is also present; this serves to regenerate the palladium chloride, as it is used up in the reaction.]

Acetone is prepared similarly from propene, but is largely made by another process involving oxidation of cumene (isopropylbenzene) which provides two important products, phenol (see Chapter 24) and acetone.

Aldehydes are also made by reaction of alkenes with carbon monoxide and hydrogen in the presence of an organocobalt catalyst. Aldehydes with one more carbon atom than the alkene are produced. An example is

$$
\begin{array}{l}
CH_3 \\
| \\
CH \\
\| \\
CH_2
\end{array}
+ CO + H_2 \xrightarrow[\substack{heat, \\ high\ pressure}]{Co_2(CO)_8}
\begin{array}{l}
CH_3 \\
| \\
CH_2 \\
| \\
CH_2 \\
| \\
CHO
\end{array}
+
\begin{array}{l}
CH_3 \\
| \\
CH-CHO \\
| \\
CH_3
\end{array}
$$

The carbon monoxide adds to either end of the double bond.

The use of organometallic compounds as catalysts in all these industrial processes should be noted.

General Chemistry of Aldehydes and Ketones; Oxidation

In general the chemical reactions of aldehydes and ketones are very similar, but ketones tend to be less reactive. This is because the presence of the second alkyl group in the case of a ketone means that the carbonyl group is more crowded by surrounding atoms than it is in the case of an aldehyde, in which there is only one large group and one small hydrogen atom. Thus approach of a reagent to the carbonyl group is more hindered in the case of a ketone, making the reaction proceed more slowly.

The only real chemical difference between aldehydes and ketones is that aldehydes are easily oxidized to carboxylic acids:

$$RCHO \xrightarrow{\text{oxidation}} RCOOH$$

This reaction is readily brought about by aqueous solutions of common oxidizing agents such as dichromates. In contrast ketones are only oxidized with more difficulty and reaction involves cleaving the carbon–carbon bond which links an alkyl group to the carbonyl group.

Reduction

The carbonyl group may be reduced to a methylene (CH_2) group or to an alcohol group, depending upon the identity of the reducing agent. Thus a mixture of zinc amalgam and hydrochloric acid reduces many ketones to hydrocarbons, but does not reduce aldehydes satisfactorily:

$$RCOR' \xrightarrow[\text{HCl}]{\text{ZnHg}} RCH_2R'$$

It is much more straightforward to reduce aldehydes or ketones to, respectively, primary or secondary alcohols:

$$RCHO \longrightarrow RCH_2OH$$
$$RCOR' \longrightarrow RCHOHR'$$

The commonest reagents are:

(a) hydrogen in the presence of a catalyst (for example Pd, Pt),
(b) sodium and ethanol (or other alcohol),
(c) sodium borohydride ($NaBH_4$),
(d) lithium aluminium hydride ($LiAlH_4$).

In reduction by sodium and an alcohol the sodium transfers an electron to the carbonyl group; this is followed by removal of a proton from the alcohol.

Sodium borohydride and lithium aluminium hydride each act as sources of hydride ions, H^-. A simplified form of the mechanism can be written as follows [the 'real' mechanism is more complicated, involving the borohydride (BH_4^-) or aluminium hydride (AlH_4^-) anions]:

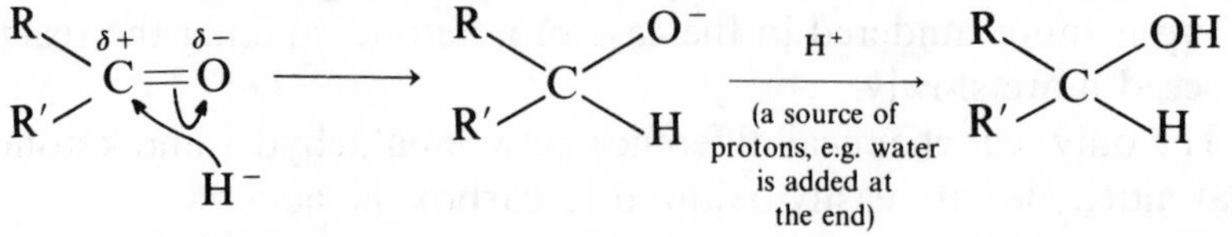

(R' = alkyl or H)

Note that this reaction involves nucleophilic attack at the carbon atom of the carbonyl group and displacement of two electrons of the carbon–oxygen double bond onto the oxygen atom. Other nucleophiles react similarly.

Reactions with Nucleophiles

The general equation for the reactions of nucleophiles with aldehydes or ketones is

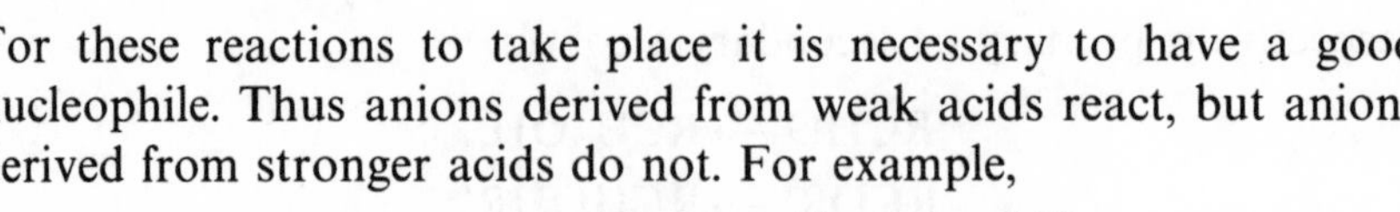

(R' = alkyl or H,
Nu$^-$ = nucleophile)

For these reactions to take place it is necessary to have a good nucleophile. Thus anions derived from weak acids react, but anions derived from stronger acids do not. For example,

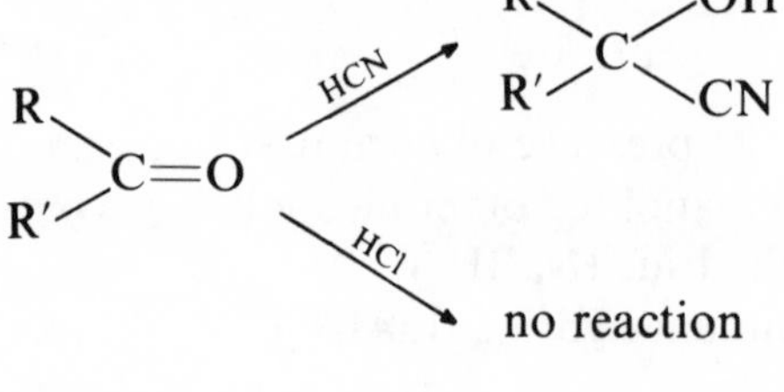

no reaction

Reactions with Hydrogen Cyanide

It is of interest that the reaction with hydrogen cyanide was one of the first organic reactions to be investigated as to its mechanism, as long ago as in 1903.

If, as we now know, the mechanism involves initial attack by a cyanide ion, that is

$$\underset{R'}{\overset{R}{\diagup}}C{=}O \xrightarrow{CN^-} \underset{R'}{\overset{R}{\diagup}}C\overset{O^-}{\underset{CN}{\diagdown}} \xrightarrow{H_2O} \underset{R'}{\overset{R}{\diagup}}C\overset{OH}{\underset{CN}{\diagdown}}$$

then the rate of reaction should depend on the concentration of cyanide ions; rate $\propto [CN^-]$.

The workers in 1903 noted the rate of reaction of camphor, a ketone, under the following sets of conditions, with the results shown:

Camphor + HCN	Reaction complete in 10 min
Camphor + HCN + dilute mineral acid	No reaction
Camphor + HCN + aqueous alkali	Reaction complete in seconds

Hydrogen cyanide is a weak acid:

$$H{-}CN \rightleftharpoons H^+ + CN^-$$

The presence of a mineral acid suppresses this ionization and in consequence very few cyanide ions are present. Alkali, on the other hand, increases the concentration of cyanide ions because of the reaction

$$HCN + HO^- \rightleftharpoons H_2O + CN^-$$

Therefore the above results show that the greater the number of cyanide ions present, the faster reaction proceeds, and that the rate is **not** proportional to the concentration of hydrogen ions present.

The products from this reaction are called **cyanhydrins**. The **cyano** group is readily converted into a carboxylic acid group, so this provides a means of making an α-hydroxycarboxylic acid:

$$\underset{R'}{\overset{R}{\diagup}}C{=}O \xrightarrow{HCN} \underset{R'}{\overset{R}{\diagup}}C\overset{OH}{\underset{CN}{\diagdown}} \xrightarrow[H^+]{H_2O} \underset{R'}{\overset{R}{\diagup}}C\overset{OH}{\underset{COOH}{\diagdown}}$$

Reactions with Water

Carbonyl groups react readily with hydroxide ions to form dihydroxy compounds but, with rare exceptions, these exist only in aqueous solution, and when water is removed, the carbonyl compound is reformed:

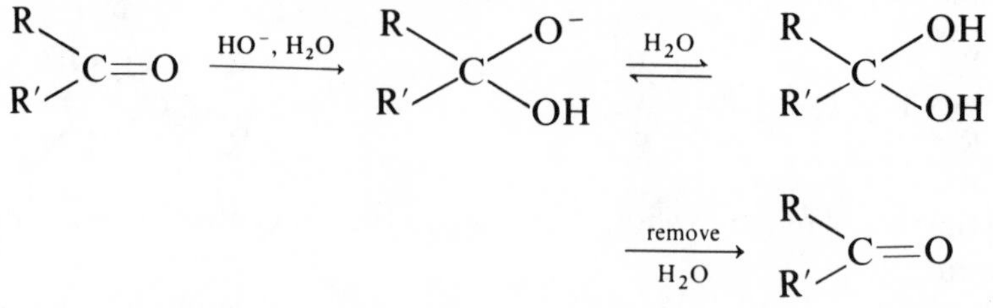

When a carbonyl compound is dissolved in water, a dihydroxy compound can be generated. Removal of the water completely reverses the equilibria involved and the carbonyl compound reappears:

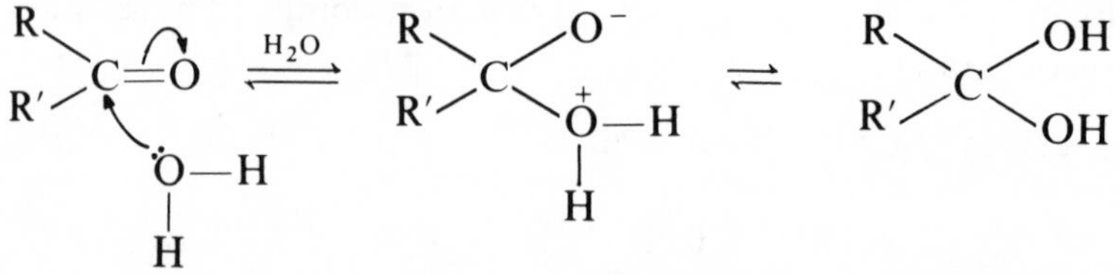

Very few organic compounds with two hydroxy groups attached to the same carbon atom retain this structure when they are isolated from an aqueous solution. A rare example is chloral hydrate, $Cl_3CCH(OH)_2$, derived from chloral, or trichloroacetaldehyde, Cl_3CCHO. The neighbouring chlorine atoms serve to stabilize the diol. This is another example of the effect that adjacent functional groups can have on one another.

Reactions with Derivatives of Ammonia

Ammonia is a stronger nucleophile than water and reacts more readily than water with carbonyl compounds. Aldehydes react more readily than ketones. The adducts formed commonly react further to provide more complex products:

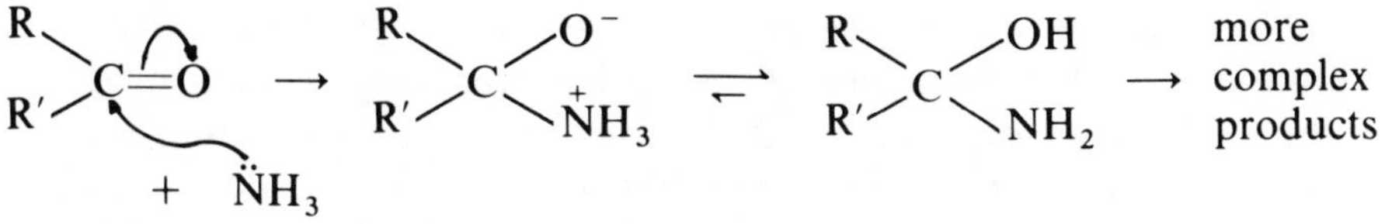

$$R_{R'} \!\!>\!\! C\!\!=\!\!O \ + \ \ddot{N}H_3 \quad \longrightarrow \quad R_{R'}\!\!>\!\!\underset{\overset{+}{N}H_3}{\overset{O^-}{C}} \quad \rightleftharpoons \quad R_{R'}\!\!>\!\!\underset{NH_2}{\overset{OH}{C}} \quad \longrightarrow \quad \begin{array}{l}\text{more}\\\text{complex}\\\text{products}\end{array}$$

Thus formaldehyde reacts with ammonia to give hexamethylenetetramine, or, for short, hexamine, $(CH_2)_6N_4$. This molecule has an interesting molecular shape:

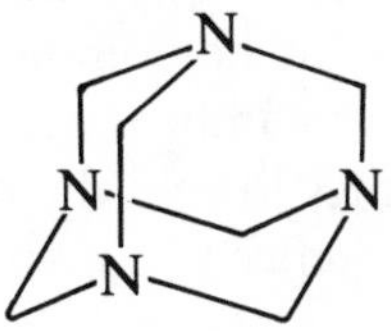

It has been used as a urinary antiseptic, and finds considerable use in the manufacture of plastics. It provides a convenient way of handling and transporting formaldehyde, which is a gas, and is otherwise transported in aqueous solution, since it is a solid and can readily be reconverted into formaldehyde. The powerful high explosive RDX is made by the reaction of hexamine with nitric acid.

Derivatives of ammonia, NH_2X, where X = OH, NH_2, etc., undergo useful reactions with carbonyl compounds. Their general form is

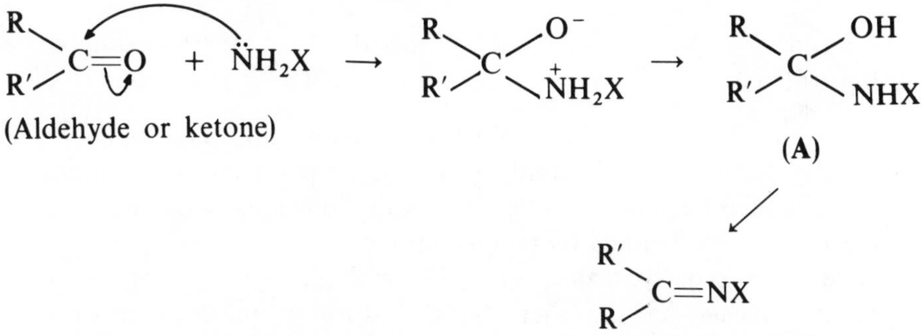

$$R_{R'}\!\!>\!\! C\!\!=\!\!O \ + \ \ddot{N}H_2X \quad \longrightarrow \quad R_{R'}\!\!>\!\!\underset{\overset{+}{N}H_2X}{\overset{O^-}{C}} \quad \longrightarrow \quad R_{R'}\!\!>\!\!\underset{NHX}{\overset{OH}{C}}$$

(Aldehyde or ketone)

$$\textbf{(A)}$$

$$\underset{R}{\overset{R'}{>}}C\!\!=\!\!NX$$

The addition product (A) loses a molecule of water to give a compound having a C=N double bond. The reaction can be summarized as

$$R_2CO + NH_2X \longrightarrow R_2C = NX + H_2O$$

An overall reaction of this sort, in which two molecules join together and in the process a small molecule, such as water or ammonia, is lost, is known as a **condensation reaction**.

The reactions of aldehydes or ketones with a variety of derivatives of ammonia, NH_2X, have proved useful for preparing crystalline products which are described as **derivatives** of the carbonyl compound from which they are obtained. Identification of unknown compounds depends on recording their physical properties. Nowadays this means recording spectra that can provide most, if not all, of the information needed. Before the ready availability of spectrometers, determination of melting points played a crucial role; even now a melting point is an important characteristic of a compound that needs to be determined. Not only can the melting point be compared with that of a known specimen, but it also provides an indication of the purity of the sample, since the presence of impurities lowers the melting point from that of the pure sample. It was the practice to convert liquid samples into solid **derivatives** so that a melting point of that derivative could be recorded and compared with the melting point of similar derivatives made from samples of known compounds. Various compounds of general structure NH_2X have been used extensively to make such derivatives from liquid aldehydes and ketones, since they frequently provide crystalline products. For example, hydroxylamine, NH_2OH, provides derivatives known as **oximes**. An example is

(Liquid,   (Solid,   Cyclohexanone
b.p. 156 °C)   m.p. 91 °C)   oxime

Now that X-ray crystallography (see the Appendix) is becoming an everyday technique, a different use of the preparation of crystalline derivatives has become valuable, to provide from a liquid compound a solid that can be studied by this technique.

The reaction of hydrazine, NH_2NH_2, with aldehydes or ketones to give **hydrazones** has another useful feature in that, when these derivatives are heated in strongly basic conditions, they are converted into alkanes:

$$R_2CO + NH_2NH_2 \rightarrow R_2C{=}NNH_2 \xrightarrow[\text{heat}]{\text{base}} R_2CH_2$$

A hydrazone

This thus provides an alternative way of reducing a carbonyl group in an aldehyde or ketone to a methylene group.

Reaction with Sodium Hydrogen Sulphite

Aldehydes and some, but not all, ketones react with sodium hydrogen sulphite as follows:

$$RCHO \;+\; Na^+\;HSO_3^- \;\rightleftharpoons\; RCH\!\!\begin{array}{l} \nearrow O^- \\ \searrow SO_3H \end{array} \;\rightleftharpoons\; RCH\!\!\begin{array}{l} \nearrow OH \\ \searrow SO_3^- \end{array}$$
$$\qquad\qquad\qquad\qquad\qquad\qquad Na^+ \qquad\qquad\qquad Na^+$$

The value of this reaction again lies in the fact that it frequently provides crystalline addition products. These may be filtered off and easily reconverted into the aldehyde (or ketone) by treatment with dilute aqueous acid or base. This has frequently provided a useful method for separating aldehydes from a mixture of compounds, such as may be obtained from natural sources, e.g. the extract of a plant, since the product that the aldehyde gives can, as a solid, be filtered off from the other materials present.

Reaction with Grignard Reagents

As has been seen in the previous chapter, Grignard reagents react with formaldehyde or other aldehydes or ketones to give, respectively, primary, secondary or tertiary alcohols:

$$RMgBr \begin{array}{l} \xrightarrow[\;(2)\,H^+,\,H_2O\;]{\;(1)\,HCHO\;} RCH_2OH \\[2mm] \xrightarrow[\;(2)\,H^+,\,H_2O\;]{\;(1)\,R'CHO\;} RCHOHR' \\[2mm] \xrightarrow[\;(2)\,H^+,\,H_2O\;]{\;(1)\,R'R''CO\;} R\!\!-\!\!\overset{\displaystyle R'}{\underset{\displaystyle R''}{C}}\!\!-\!\!OH \end{array}$$

Reactions with Electrophiles

Since the positively charged carbon atom of a carbonyl group reacts with nucleophiles, it is reasonable to consider whether the negatively charged oxygen atom can react with electrophiles. The only general example of such reactivity is that of aldehydes or ketones with protons, i.e. with protic acids:

$$\underset{R'}{\overset{R}{>}}C=O \quad H^+ \quad \longrightarrow \quad \underset{R'}{\overset{R}{>}}\overset{+}{C}-OH$$

Two of the electrons of the double bond interact with the proton to form an O—H bond, thus leaving the carbon atom deficient in electrons, i.e. positively charged.

However, the adjacent oxygen atom can help to share this positive charge by sharing a lone pair of electrons with the carbon atom. This can be written as

$$\underset{R'}{\overset{R}{>}}\overset{+}{C}-\overset{..}{O}-H \quad \longleftrightarrow \quad \underset{R'}{\overset{R}{>}}C=\overset{+}{O}-H$$

$$\textbf{(B)} \qquad\qquad\qquad \textbf{(C)}$$

If a full double bond were formed the oxygen atom, and not the carbon atom, would bear the full positive charge. In fact the pair of electrons are only partially shared between the carbon and oxygen atoms, and the real structure is partway between structure **B** and structure **C**, or alternatively can be thought of as an average of structures **B** and **C**. This is the significance of the doubly-headed arrow. It does **not** mean that structures **B** and **C** are continuously interchanging but that the overall structure is a mean of the structures involved. In this way of describing the compound it is called a **resonance hybrid** of forms **B** and **C**. Another way of representing the situation is

$$\underset{R'}{\overset{R}{>}}C\overset{+}{-\!-\!-}O-H$$

The dotted line shows that the double bond is only partially formed and the positive charge is shared between the carbon and oxygen atoms. This type of dotted-line formula will be commonly used in this book. The dotted line is said to show the **delocalization** of two electrons, in this case between the oxygen atom and the bond linking the carbon and oxygen atoms, and of positive charge, in this case between the carbon and oxygen atoms.

If charge can be shared, or delocalized, between atoms, it helps to increase the stability of the molecule, compared to a situation in which a charge is perforce located, or localized, on one atom alone. A further example of this will be met later in this chapter, and in the following chapter.

The salts so formed from carbonyl compounds and acids cannot normally be isolated. They may undergo further reaction with nucleophiles, reaction taking place at the partially positively charged carbon atom. Thus in aqueous solution a dihydroxy compound, or **diol**, may be formed:

$$\underset{R'}{\overset{R}{>}}C=O \xrightarrow{\;H^+\;} \underset{R'}{\overset{R}{>}}C\overset{-\cdots+}{-}OH \rightleftharpoons \underset{R'}{\overset{R}{>}}C\underset{\overset{+}{O}-H}{<}OH$$

$$\overset{\ddot{O}-H}{\underset{H}{|}}$$

$$\rightleftharpoons^{\;-H^+\;} \underset{R'}{\overset{R}{>}}C\underset{OH}{<}OH$$

In aqueous solution many carbonyl compounds exist as diols. (Water also provides the proton for the first step.) As mentioned above, in most cases, on attempted isolation from solution the diols revert to carbonyl compounds:

$$R_2C(OH)_2 \rightarrow R_2CO + H_2O$$

Acetals and Ketals

If, on the other hand, a solution of a carbonyl compound in ethanol is treated with acid, isolable products are obtained—**acetals** from aldehydes and **ketals** from ketones:

$$\underset{H}{\overset{R}{>}}C=O \xrightarrow[C_2H_5OH]{H^+,} \underset{H}{\overset{R}{>}}C\underset{OC_2H_5}{<}OC_2H_5 \qquad \text{An acetal}$$

$$\underset{R'}{\overset{R}{>}}C=O \xrightarrow[C_2H_5OH]{H^+,} \underset{R'}{\overset{R}{>}}C\underset{OC_2H_5}{<}OC_2H_5 \qquad \text{A ketal}$$

In this case the products are diethers rather than diols. Other alcohols react similarly. As for a number of reactions of aldehydes and ketones, reaction proceeds less readily in the case of ketones.

The mechanism for their formation is as follows:

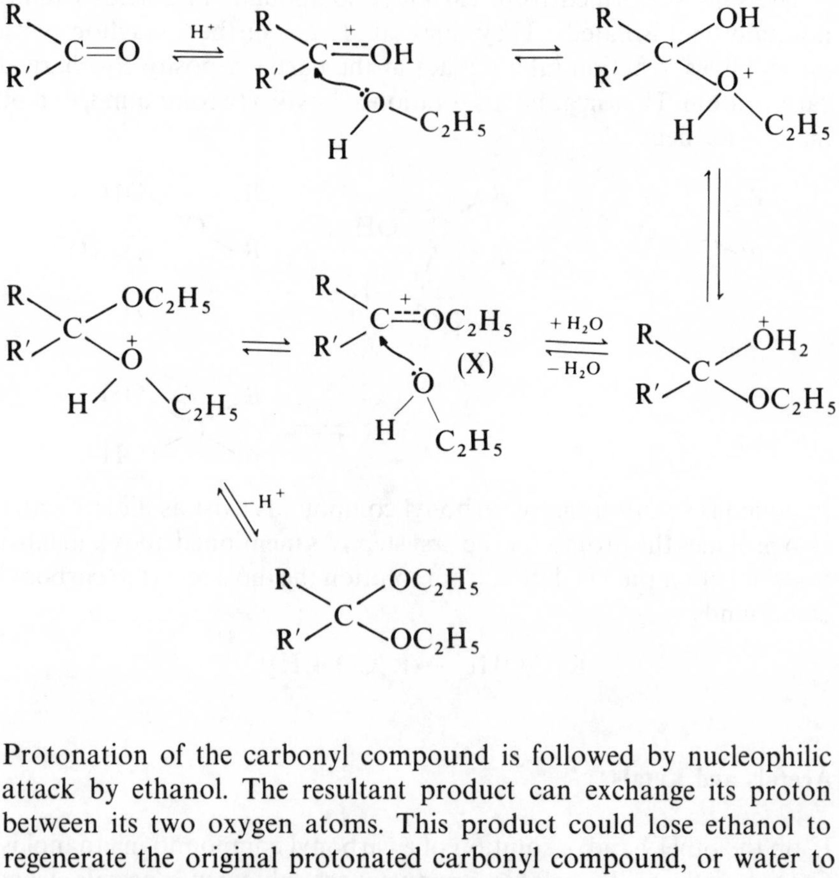

Protonation of the carbonyl compound is followed by nucleophilic attack by ethanol. The resultant product can exchange its proton between its two oxygen atoms. This product could lose ethanol to regenerate the original protonated carbonyl compound, or water to form a different protonated species (X), which can react with more ethanol to provide the acetal or ketal. Note that all these steps are equilibria and are reversible. Excess of ethanol, present as solvent, drives the equilibria, and hence the overall reaction, towards formation of the acetal or ketal.

However, if the acetal or ketal is dissolved in aqueous acid, water is present in excess and the whole process is reversed, and the carbonyl compound is regenerated.

The situation can be summed up by the equilibrium

$$\underset{R'}{\overset{R}{\diagdown}}C=O \quad \underset{H_2O,\,H^+}{\overset{C_2H_5OH,\,H^+}{\rightleftharpoons}} \quad \underset{R'}{\overset{R}{\diagdown}}C\underset{\diagdown OC_2H_5}{\diagup OC_2H_5}$$

Acetals and ketals are not decomposed in alkaline solution.

Acetals and Ketals as Protecting Groups

This equilibrium is made use of to 'protect' carbonyl groups in molecules while reactions are carried out on another part of the molecule.

Suppose that there was need to convert a bromoaldehyde such as $BrCH_2CH_2CH_2CHO$ into a cyanoaldehyde $NCCH_2CH_2CH_2CHO$. If the aldehyde was treated with cyanide, this might react with the carbonyl group as well as substituting the bromine atom. If, however, the aldehyde group is first converted into an acetal, the latter grouping does not react with nucleophiles. Then the acetal can react with cyanide to give a cyanoacetal; treatment of this product with dilute aqueous acid provides the required cyanoaldehyde (conversion of the acetal into the aldehyde would need to be done gently, i.e. using cold dilute acid, for stronger acid conditions could also convert the cyano group into a carboxylic acid group).

$$BrCH_2CH_2CH_2CHO \xrightarrow{C_2H_5OH,\,H^+} BrCH_2CH_2CH_2CH(OC_2H_5)_2$$

$$\downarrow KCN$$

$$NCCH_2CH_2CH_2CHO \xleftarrow{H_2O,\,H^+} NCCH_2CH_2CH_2CH(OC_2H_5)_2$$

Thioacetals and Thioketals

Aldehydes and ketones can similarly be converted into thioacetals or thioketals, if treated with thiols in the presence of acid. Again dilute aqueous acid regenerates the carbonyl compound:

$$\underset{R'}{\overset{R}{>}}C=O \quad \underset{\overset{H^+}{\longleftarrow}}{\overset{C_2H_5SH,\ H^+}{\longrightarrow}} \quad \underset{R'}{\overset{R}{>}}C\underset{SC_2H_5}{\overset{SC_2H_5}{<}}$$

Acid Catalysis of Reactions of Nucleophiles with Carbonyl Compounds

Acids can catalyse the addition reactions of nucleophiles to carbonyl compounds. Protonation of the carbonyl group results in its bearing a positive charge and it is thus more reactive towards nucleophiles:

$$\underset{R'}{\overset{R}{>}}C=O \quad \overset{H^+}{\longrightarrow} \quad \underset{R'}{\overset{R}{>}}\overset{+}{C}\text{---}OH \quad \longrightarrow \quad \underset{R'}{\overset{R}{>}}C\underset{Nu}{\overset{OH}{<}}$$

with Nu^- attacking

$$(Nu^- = \text{nucleophile})$$

For example, reactions involving derivatives of ammonia, NH_2X, are assisted in weakly acidic conditions.

$$\underset{R'}{\overset{R}{>}}C=O \quad \overset{H^+}{\longrightarrow} \quad \underset{R'}{\overset{R}{>}}\overset{+}{C}\text{---}OH \quad \longrightarrow \quad \underset{R'}{\overset{R}{>}}C\underset{\overset{+}{N}H_2X}{\overset{OH}{<}}$$

with $:NH_2X$ attacking

$$\Big\Vert -H^+$$

$$\underset{R'}{\overset{R}{>}}C=NX \quad \overset{-H_2O}{\longleftarrow} \quad \underset{R'}{\overset{R}{>}}C\underset{NHX}{\overset{OH}{<}}$$

It is important that the conditions are not too strongly acidic, for if this is the case the nucleophile will be protonated:

$$:NH_2X \overset{H^+}{\longrightarrow} \overset{+}{N}H_3X$$

This protonated species is not a nucleophile because the lone pair of electrons has already been used up to form the ammonium salt.

The optimum pH for acid catalysed reactions of carbonyl compounds with such nucleophiles is ~ 5.

Acidity of α-Hydrogen Atoms in Aldehydes and Ketones

The presence of a carbonyl group affects the properties of adjacent C—H bonds, and, in consequence, α-hydrogen atoms, i.e. hydrogen atoms attached to carbon atoms next to the carbonyl group, are weakly acidic. They are removed as protons by strong bases:

$$\underset{\overset{\displaystyle |}{\underset{\displaystyle \overset{|}{H}}{H}}}{\overset{\displaystyle \overset{|}{H}}{R-C-COR}} \longrightarrow R-\underset{|}{\overset{|}{C}}-COR + H_2O$$

$$\overset{\displaystyle H \curvearrowleft}{\underset{\displaystyle \overline{O}H}{}}$$

The resultant anion is stabilized, because the negative charge can be shared between the α-carbon atom and the oxygen atom and, as discussed earlier in this chapter, sharing, or delocalization, of charge stabilizes an ion. This delocalization can be represented either by the resonance hybrid picture

$$R-\underset{|}{\overset{\displaystyle \overset{H}{|}}{C}}-\overset{\displaystyle \overset{O}{\parallel}}{C}-R \longleftrightarrow R-\overset{\displaystyle \overset{H}{|}}{C}=\overset{\displaystyle \overset{O^-}{|}}{C}-R$$

or by the dotted line symbolism

$$R-\underset{\delta-}{\overset{\displaystyle \overset{H}{|}}{C}}\overset{\displaystyle \overset{O^{\delta-}}{\vdots}}{C}-R$$

In the latter the δ-signs at C and at O indicate that the negative charge is shared between these two atoms.

Another feature brought out by both of these representations is that, in consequence of this delocalization, the carbonyl group has less than full double-bond character, while some alkene double-bond character is present in the adjacent carbon–carbon bond.

The formation of such ions leads to a number of reactions which take place involving α-carbon atoms of carbonyl compounds. Two examples are given.

Halogenation of Carbonyl Compounds in the Presence of Base

An example of such a reaction is

$$CH_3CHO \xrightarrow[\text{NaOH}]{\text{Br}_2,} CH_2BrCHO$$

The reaction involves formation of an anion which then reacts with bromine:

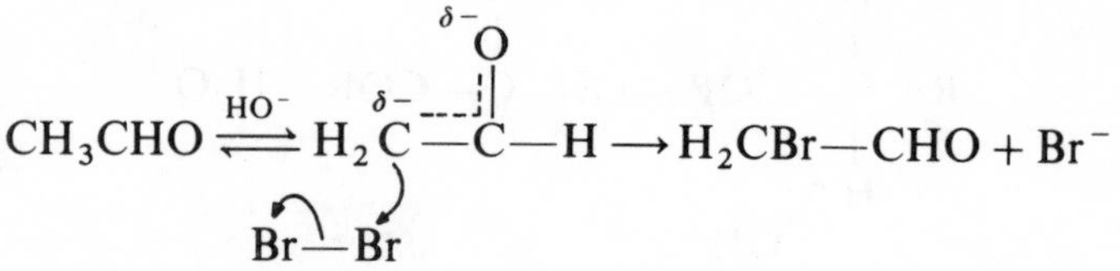

Reaction takes place at carbon rather than at oxygen because O—Br bonds are very weak whereas C—Br bonds are much stronger. Chlorine reacts similarly. Any aldehydes or ketones having hydrogen atoms attached to the carbon atoms next to the carbonyl group can take part in such reactions. Another example is

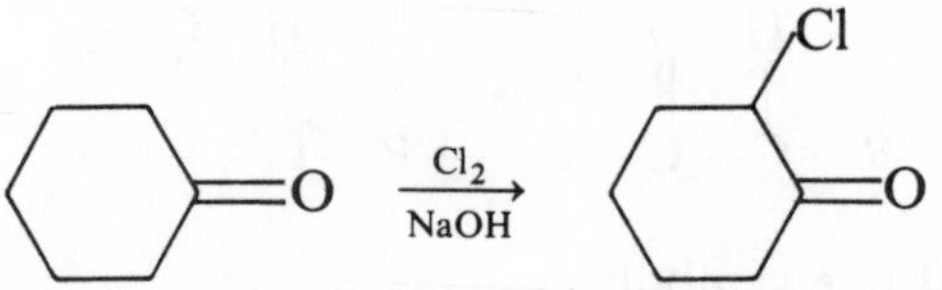

More than one α-hydrogen atom may be replaced if sufficient bromine or chlorine is available. Thus acetaldehyde can be converted into trichloroacetaldehyde (chloral)

$$CH_3CHO \xrightarrow[\text{NaOH}]{3\,\text{Cl}_2,} CCl_3CHO$$

Self-condensation Reactions

When acetaldehyde is treated with a strong base the following reactions ensue:

$$CH_3CHO + CH_3CHO \xrightarrow[\text{cold}]{\text{NaOH}} CH_3 - \underset{\underset{OH}{|}}{\overset{\overset{H}{|}}{C}} - CH_2CHO \xrightarrow[\text{heat}]{\text{NaOH}}$$

$$CH_3CH = CHCHO + H_2O$$

Similarly two acetone molecules can react together:

$$2\,CH_3COCH_3 \xrightarrow[\text{heat}]{\text{NaOH}} \quad \overset{CH_3}{\underset{CH_3}{}}{>}C = CHCOCH_3 + H_2O$$

In each case the net result is a condensation reaction between two molecules of the aldehyde or ketone.

For such a self-condensation reaction to be possible, at least one of the carbons next to the carbonyl group must have two hydrogen atoms attached to it, which can be lost in the condensation process.

Reaction proceeds via formation of an anion which then reacts with another molecule of aldehyde or ketone. The mechanism is

The final elimination reaction takes place readily because of the acidity of the α-hydrogen atom, which is removed by a hydroxide ion.

Note that this condensation process could continue:

condensation polymer ◄——————— etc.

Many molecules of acetaldehyde can link together, with loss of water as each additional acetaldehyde molecule reacts, providing large molecules which are not strictly polymers of the aldehyde, which would have the formula $(CH_3CHO)_n$, but instead are represented by $[(CH_3CHO)_n - (n + 1)H_2O]$. A process like this is known as **condensation polymerization** and the products are **condensation polymers**. Such reactions are possible for all carbonyl compounds, aldehydes and ketones, which have α-methylene (CH_2) groups.

Questions

1. A compound C_4H_8O shows a carbonyl peak in its infrared spectrum. What are the possible structural formulae? How could you distinguish between these possibilities by means of ^1H-n.m.r. spectra? Show by equations how you might make **one** of the possible isomers starting from 1-chlorobutane.

2. Show by means of an equation the product that would arise from the reaction of acetone with hydroxylamine (NH_2OH). Give the mechanism of the reaction.

3. Why do ketones react with HCN in alkaline solution but not in acidic solution?

4. Show by means of equations what reagents you would use to convert acetone into (a) propan-2-ol, (b) $(CH_3)_2C{=}NNHCONH_2$ and (c) 2-methylpropan-2-ol.

5. What sequences of reactions would you use to convert (a) ethanol into $CH_3CH{=}CHCHO$, (b) ethanol into propan-2-ol and (c) 4-bromobutanal into 4-hydroxybutanal?

6. Which of the following compounds will react with (i) NaCN and (ii) HBr?

(a) $CH_3CHClCH_2CH_3$,   (b) $CH_3COCH_2CH_3$,
(c) $(CH_3)_2CHCH_2CH_3$,   (d) $CH_3CHOHCH_2CH_3$,
(e) $(CH_3)_2CHCH{=}CH_2$,   (f) $CH_3CHNH_2CH_2CH_3$.

Write an equation for each reaction, using curved arrows where appropriate to show the mechanism. Indicate whether the initial attack on the compounds (a) to (f) is by an electrophile or a nucleophile.

7. A compound $C_6H_{10}O$ undergoes the following series of reactions:

$$C_6H_{10}O \xrightarrow{?} C_6H_{12}O \xrightarrow{?} C_6H_{10} \xrightarrow{?} C_6H_{12}$$

$$\text{(A)} \qquad\quad \text{(B)} \qquad\quad \text{(C)} \qquad\quad \text{(D)}$$

The final product shows only one signal in both its ^1H- and ^{13}C-n.m.r. spectra. What are the reagents used for each step, and what are the structural formulae of compounds **A**, **B**, **C** and **D**?

15 CARBOXYLIC ACIDS

The final products of the oxidation of primary alcohols or aldehydes are **carboxylic acids**.

$$R-C\overset{O}{\underset{OH}{\diagup}}$$

The—COOH group is known as the **carboxyl** group.

The carboxylic group has a hydroxy group attached to the carbon atom of a carbonyl group. Since these two groups are adjacent to one another it is to be expected that they will interact with each other, and that their properties will be modified from those of isolated carbonyl or hydroxy groups.

This modification is very obvious in the properties of carboxylic acids. Thus they do not take part in the characteristic reactions of aldehydes and ketones with nucleophiles. In particular the hydroxy group of a carboxylic acid is quite strongly acidic, in contrast to the effective neutrality of alcohols.

At the end of the preceding chapter the interaction between a carbonyl group and an adjacent anionic centre was noted. A somewhat similar situation is present in a carbonyl group. An isolated carbonyl group is polarized $\overset{\delta+}{C}=\overset{\delta-}{O}$. This can be represented in resonance nomenclature as $C=O \leftrightarrow \overset{+}{C}-\overset{-}{O}$. In a carboxyl group, next to the potentially positively charged carbon atom is another oxygen atom. This atom has two lone pairs of electrons associated with it. These electrons can interact with a cationic carbon atom in the same way as happens in protonated aldehydes or ketones, namely

$$\overset{\diagdown}{\underset{\diagup}{C}}\!\!\overset{+}{-}\ddot{O}H \;\longleftrightarrow\; \overset{\diagdown}{\underset{\diagup}{C}}\!\!=\!\overset{+}{O}H \quad or \quad \overset{\diagdown}{\underset{\diagup}{C}}\!\!\overset{+}{\doteq}OH$$

If we combine this interaction with the tendency to polarize of the carbonyl group, the resultant situation can be summarized, in resonance terms, as

$$R\!-\!C\!\!\underset{OH}{\overset{O}{\diagup}} \;\longleftrightarrow\; R\!-\!\overset{+}{C}\!\!\underset{OH}{\overset{O^-}{\diagup}} \;\longleftrightarrow\; R\!-\!C\!\!\underset{\underset{+}{OH}}{\overset{O^-}{\diagup}}$$

or, using dotted line symbolism, as

$$R\!-\!C\!\!\underset{OH}{\overset{O^{\delta-}}{\diagup}}{\scriptstyle\delta+}$$

In general terms, the formally doubly bonded oxygen atom bears a partial negative charge, while a partial positive charge is shared between the other oxygen atom and the carbon atom.

This means that the carbon atom of a carboxyl group carries less positive charge than does the carbon atom of a simple carbonyl group, and is hence less reactive towards nucleophiles. The positive charge shared between the carbon and oxygen atoms tends to weaken the O—H bond, making it more acidic, although this is a partial, and not the only, contributing factor to the acidity of carboxylic acids, which will be discussed later in this chapter.

Nomenclature

The systematic name of a carboxylic acid is made by removing the final 'e' from the name of the alkane with the same number and arrangement of carbon atoms and replacing it by 'oic acid'. Some examples are as follows:

$CH_3CH_2CH_2COOH$	Butanoic acid
$CH_3CH_2CH_2CH_2COOH$	Pentanoic acid
$CH_3CH_2CHClCOOH$	2-Chlorobutanoic acid
$\underset{\displaystyle CH_3CH_2CHCH_2COOH}{\overset{\displaystyle CH_3}{\overset{\displaystyle \vert}{}}}$	3-Methylpentanoic acid

The carbon atom of the carboxyl group is always numbered 1. Sometimes letters of the Greek alphabet are used instead of numbers to indicate the position of substituent groups, as in

$CH_3CH_2CHClCOOH$ α-Chlorobutanoic acid
$CH_3CH_2CHBrCH_2COOH$ β-Bromopentanoic acid

For the lowest members of the series it is normal, and official, practice to use the trivial names:

        HCOOH   Formic acid
        CH_3COOH Acetic acid
        CH_3CH_2COOH Propionic acid

Long-established trivial names are also standardly used for some naturally occurring longer-chain carboxylic acids which are of importance, such as

$CH_3(CH_2)_{14}COOH$ Palmitic acid
                            (hexadecanoic acid)
$CH_3(CH_2)_{16}COOH$ Stearic acid
                            (octadecanoic acid)
$CH_3(CH_2)_7CH{=}CH(CH_2)_7COOH$ Oleic acid
                            (octadec-9-enoic acid)

Natural Occurrence

Carboxylic acids are widespread in nature. Formic acid is partly responsible for the irritation resulting from contact with stinging nettles, and from ant and bee stings. It was first made by distillation of red ants. This association with ants accounts for its name since *formica* is the Latin word for an ant. [Germans call it ameisensäure (*ameise* = ant; *säure* = acid), so perhaps it should be called *antic acid* in English!]. Acetic acid is a product of fermentation and arises from the souring of alcoholic materials such as wines and beers. It was once made by distillation of wood. Vinegar is a dilute aqueous solution of acetic acid. Butanoic acid, sometimes also called butyric acid, is present in rancid butter and is partly responsible for its smell. Longer-chain acids are widely distributed in vegetable and animal fats, often in the form of esters, discussed in a later chapter. These acids, for the most part, have unbranched chains of carbon atoms and even numbers of carbon atoms. They may be saturated or contain double bonds.

Preparation of Carboxylic Acids

The simpler carboxylic acids are made industrially by oxidation of the appropriate aldehyde. Formic acid is also made from carbon monoxide and sodium hydroxide:

$$CO + NaOH \xrightarrow[\text{pressure}]{\text{heat}} HCOONa \xrightarrow[\text{acid}]{\text{mineral}} HCOOH$$

$$\text{Sodium}$$
$$\text{formate}$$

It has use industrially as an inexpensive weak acid and solvent.
Laboratory methods of preparation, already discussed earlier, are

1. Oxidation of primary alcohols and aldehydes:

$$RCH_2OH \xrightarrow{\text{oxidation}} RCHO \xrightarrow{\text{oxidation}} RCOOH$$

2. From alkyl halides, via alkyl cyanides (nitriles):

$$RCl \xrightarrow{\text{NaCN}} RCN \xrightarrow{H^+, H_2O} RCOOH$$

3. From alkyl halides, via Grignard reagents:

$$RBr \xrightarrow[\substack{\text{dry ether} \\ \text{dry } N_2}]{\text{Mg}} RMgBr \xrightarrow[\text{(2) } H^+, H_2O]{\text{(1) } CO_2} RCOOH$$

Physical Properties

The lower members of the series of carboxylic acids are soluble in water. There is hydrogen bonding between the carboxyl groups and water molecules. They have higher melting points and boiling points than alcohols with the same number of carbon atoms, e.g. formic acid, b.p. 100 °C (cf. methanol, b.p. 65 °C), acetic acid, b.p. 118 °C, m.p. 17 °C (cf. ethanol, b.p. 78 °C, m.p. − 117 °C). This is connected with the fact that carboxylic acid molecules associate with one another by intermolecular hydrogen bonding:

(The dots indicate hydrogen bonding.)

The most striking physical characteristic of carboxylic acids, which distinguishes them from alcohols, is their acidity:

$$CH_3COOH + H_2O \rightleftharpoons CH_3CO_2^- + H_3O^+$$

The dissociation constant, K_a, of acetic acid is

$$K_a = \frac{[H_3O]^+[CH_3CO_2^-]}{[CH_3COOH]} = 1.8 \times 10^{-5}$$

This may be compared with the values for water, 2×10^{-16}, or ethanol, 10^{-18}. It indicates that in an aqueous solution of acetic acid about one molecule in 200 is dissociated into ions.

Why is it so much stronger an acid than water or an alcohol? One matter was touched on above—the weaker nature of the O—H bond. The other important factor is the stability of the **carboxylate** anion. When a proton is removed from a carboxyl group the resultant anion can be represented formally as

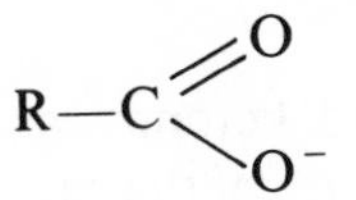

However, we know that a C=O group is polarized. The negatively charged anion pictured above could interact with this carbonyl group as follows:

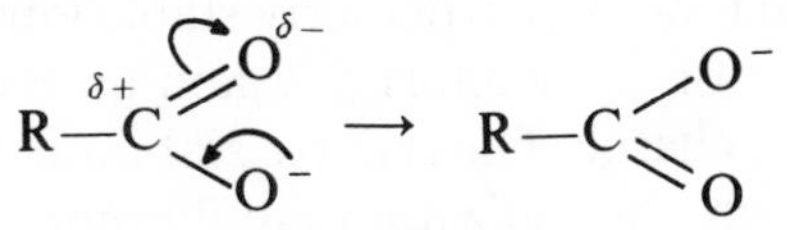

Obviously a reverse interaction is also possible:

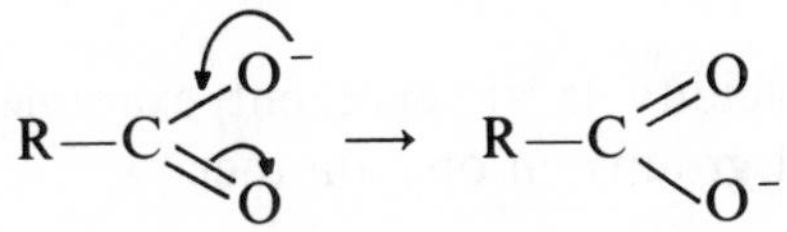

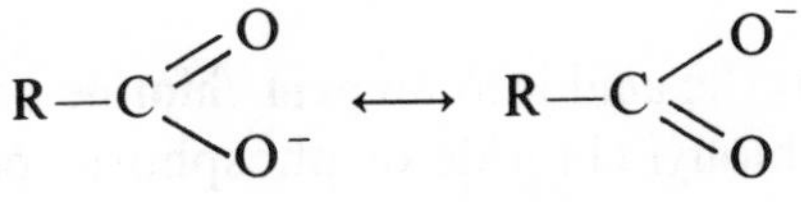

In fact the anion exists as an average of these structures, which can be expressed in resonance terms as

or in terms of dotted lines as

$$R—C\underset{O^{\delta-}}{\overset{O^{\delta-}}{\lesseqgtr}}$$

The latter representation clearly shows the symmetry of the carboxylate anion. The two oxygen atoms are equivalent, the double bond is shared equally between the two C—O bonds and the negative charge is shared equally between the two oxygen atoms. The position is very similar to that in anions

$$—\overset{|}{\underset{|}{\bar{C}H}}—\overset{O}{\overset{\|}{C}}—$$

derived from aldehydes and ketones, as discussed in the previous chapter, except that, in the case of the carboxylate anion, the charge is shared between two identical atoms and the delocalized structure is symmetrical. It was also mentioned in the previous chapter that delocalization of charge serves to stabilize an ion. This is particularly true if it is delocalized over identical atoms.

Thus carboxylate ions are particularly stable ions, whereas the ions derived from ionization of alcohols can gain no such stabilization by delocalization of the charge. The relative stabilization of the carboxylate ion in turn affects the ionization equilibrium so as to favour its formation, thus increasing K_a, compared to that of ethanol. In consequence carboxylic acids are markedly more acidic than are alcohols.

This increased acidity is an excellent example of the effects of adjacent functional groups on one another.

Replacement of Hydroxy Group by a Chlorine Atom; Formation of Acid Chlorides

To convert a carboxylic acid into an **acid chloride** or **acyl chloride** the reagents used are thionyl chloride or phosphorus pentachloride:

$$RCOOH \quad \overset{SOCl_2}{\longrightarrow} \quad RC\overset{O}{\underset{Cl}{\diagup}} \quad + SO_2 + HCl$$

$$\overset{PCl_5}{\longrightarrow} \quad RC\overset{O}{\underset{Cl}{\diagup}} \quad + POCl_3 + HCl$$

An **acid chloride** or **acyl chloride**

The former is the more convenient method, since the other products formed are both gaseous, and hence easier to separate from the acid chloride. This name draws attention to the relationship of these compounds to carboxylic acids. The grouping

is called an **acyl** group; hence the alternative name, acyl chloride. Names of individual acid chlorides are derived from the related carboxylic acid, replacing '-ic acid' by '-yl chloride', for example

CH_3COOH CH_3COCl
Acetic acid   Acetyl chloride
$CH_3(CH_2)_4COOH$ $CH_3(CH_2)_4COCl$
Hexanoic acid   Hexanoyl chloride

Reactions of Carboxylic Acids with Derivatives of Ammonia

In the preceding chapter the reactions of carbonyl compounds with derivatives of ammonia, NH_2X, to give products, $RR'C{=}NX$, were discussed. Carboxylic acids react in a totally different way with these reagents, to give ammonium salts:

$$RCOOH + NH_2X \rightarrow RCO_2^- \overset{+}{N}H_3X$$

This results from both the lower reactivity of the carboxyl–carbonyl group towards nucleophiles, and especially the acidity of the carboxyl–hydroxy group.

Reactions with Alcohols; Formation of Esters

Alcohols are far weaker bases than amines, but are still reasonable nucleophiles. Hence reaction at the carbonyl group of a carboxylic acid can compete successfully with attack at the hydrogen atom.

When an alcohol and a carboxylic acid are mixed an equilibrium is set up:

$$R-C\overset{O}{\underset{OH}{\diagup}} + R'OH \rightleftharpoons R-C\overset{O}{\underset{OR'}{\diagup}} + H_2O$$

Acid + Alcohol $\rightleftharpoons$ Ester + Water

A compound in which the acidic hydrogen atom of a carboxyl group has been replaced by an alkyl group is known as an **ester**.

At first sight this equation looks very similar to the well-known reaction: acid plus base gives salt plus water, for example,

$$CH_3COOH + NaOH \longrightarrow CH_3CO_2^- \, Na^+ + H_2O$$

Sodium acetate

In fact the two reactions are very different. Salt formation is a very fast reaction; ester formation is a fairly slow reaction. In particular, salt formation is a 'one-way' reaction, as expressed by the one-way arrow in the relevant equation. Ester formation involves setting up an equilibrium between the two reactants and the two products. At equilibrium the rates of the forward and backward reactions are identical, and the ratios of the amounts of products and initial reactants can be expressed by an equilibrium constant, K. For example, for the reaction between acetic acid and ethanol,

$$K = \frac{[CH_3COOC_2H_5][H_2O]}{[CH_3COOH][C_2H_5OH]}$$

([x] = concentration of x)

At 25 °C this contant has a value of 4. The equilibrium is only very slowly established.

Different mechanisms obtain for salt formation and ester formation. In salt formation the base removes a proton from the acid, for example

$$CH_3C\overset{O}{\underset{O-H}{\diagup}} \quad \overset{\frown}{OH} \;\; Na^+ \quad \longrightarrow \quad CH_3CO_2^- \;\; Na^+ \; + \; H_2O$$

The water that is formed provides very few protons that could bring about any reverse reaction, and salt formation is effectively irreversible.

In ester formation a sequence of reactions is involved, as shown in the following example:

$$CH_3-\overset{\overset{\displaystyle O}{\|}}{\underset{\underset{\displaystyle H}{\diagup}\overset{\displaystyle \ddot{O}}{\diagdown}C_2H_5}{C}}-OH \quad \rightleftharpoons \quad CH_3-\overset{\overset{\displaystyle O^-}{|}}{\underset{\underset{\displaystyle H}{\diagup}\overset{\displaystyle \overset{+}{O}}{\diagdown}C_2H_5}{C}}-OH \quad \rightleftharpoons \quad CH_3-\overset{\overset{\displaystyle \bar{O}}{|}}{\underset{\underset{\displaystyle}{O}\diagdown C_2H_5}{C}}-\overset{+}{O}H_2$$

$$CH_3C\overset{\displaystyle O}{\underset{\displaystyle OC_2H_5}{\diagup\diagdown}} \quad + \quad H_2O$$

Note that, since the equilibrium constant,

$$K = \frac{[\text{ester}][H_2O]}{[\text{acid}][\text{alcohol}]}$$

the presence of a large excess of alcohol leads to the formation of more ester, so that K remains constant. Hence an excess of one of the reagents, commonly the alcohol, is used to increase the yield of ester.

The reaction is catalysed and speeded up by the presence of a mineral acid. This serves to protonate the carboxylic acid molecules and make them more susceptible to nucleophilic attack by the alcohol:

$$CH_3C\overset{\displaystyle O}{\underset{\displaystyle OH}{\diagup\diagdown}} \xrightleftharpoons{H^+} CH_3C\overset{\displaystyle OH}{\underset{\displaystyle OH}{\overset{+}{\diagup}\diagdown}} \rightleftharpoons CH_3-\overset{\overset{\displaystyle OH}{|}}{\underset{\underset{\displaystyle C_2H_5\overset{+}{O}H}{|}}{C}}-OH \rightleftharpoons CH_3-\overset{\overset{\displaystyle OH}{|}}{\underset{\underset{\displaystyle OC_2H_5}{|}}{C}}-\overset{+}{O}H_2$$

$$C_2H_5\ddot{O}H$$

$$H^+ \quad + \quad CH_3C\overset{\displaystyle O}{\underset{\displaystyle OC_2H_5}{\diagup\diagdown}} \quad \rightleftharpoons \quad CH_3-\overset{\overset{\displaystyle O-H}{|}}{\underset{\underset{\displaystyle OC_2H_5}{|}}{C}}+ \quad + \quad H_2O$$

If the catalyst is concentrated sulphuric acid, this can further help ester formation by removing the water that is formed in the reaction, and thus preventing the reverse reaction from taking place.

The reverse reaction, i.e. conversion of an ester into an acid and an alcohol, can be brought about by having an excess of water present, together with a mineral acid as a catalyst. This **hydrolysis** of an ester is discussed further in Chapter 17.

Oxidation and Reduction

In general carboxylic acids are not easily oxidized, but formic acid is exceptional and is readily oxidized to carbon dioxide:

$$H-C\underset{\text{OH}}{\overset{O}{\diagup\diagdown}} \xrightarrow{\text{oxidation}} H_2O + CO_2$$

Note that formic acid, unlike other carboxylic acids, has the characteristic grouping

$$H-C\overset{O}{\underset{}{\diagup}}$$

in its structure, and aldehydes are readily oxidized.

Most of the methods of reduction described for aldehydes and ketones are ineffective for carboxylic acids. They are, however, reduced to alcohols when treated with lithium aluminium hydride, followed by water:

$$RCOOH \xrightarrow[\text{(2)}H_2O,H^+]{\text{(1)}LiAlH_4} RCH_2OH$$

Dicarboxylic Acids

There is a series of dicarboxylic acids of general formula $HOOC(CH_2)_nCOOH$, many of which occur naturally.

Oxalic acid, $HOOC-COOH$, occurs in the cell sap of many plants, usually as a salt. The acid is very poisonous, but is normally removed by cooking from foods that contain it. It is present, for example, in rhubarb, especially in the leaves. It is made commercially by heating sodium formate:

$$2\,HCO_2^-\,Na^+ \xrightarrow{\text{heat}} Na^+\,{}^-O_2C{-}CO_2^-\,Na^+$$

Sodium oxalate

Sodium oxalate can be converted into oxalic acid by adding mineral acid. Unlike the majority of carboxylic acids, oxalic acid is easily oxidized, e.g. by cold aqueous potassium permanganate:

$$HOOC{-}COOH \xrightarrow{\text{KMnO}_4} 2\,CO_2 + H_2O$$

It is thus a reducing agent, and it has been used in this way as a stain remover, particularly for ink and iron stains. The latter are removed by converting the coloured insoluble iron(III) salts into almost colourless soluble iron(II) salts. It is also used in the bleaching of straw and of wood, and in printing and dyeing of cloth, as a mordant. When heated, oxalic acid, again unlike most carboxylic acids, decomposes, first to formic acid, which itself decomposes giving carbon monoxide and water:

$$HOOC{-}COOH \xrightarrow{\text{heat}} CO_2 + HCOOH \xrightarrow{\text{heat}} CO + H_2O$$

The bond linking the two carboxyl groups is weaker than most carbon–carbon bonds.

Malonic acid, $HOOCCH_2COOH$, also decomposes when heated, to give carbon dioxide and acetic acid:

$$HOOCCH_2COOH \xrightarrow{\text{heat}} CO_2 + CH_3COOH$$

This behaviour is characteristic of all dicarboxylic acids with two carboxyl groups attached to one and the same carbon atom. Another example is

$$
\begin{array}{c}
\text{COOH} \\
| \\
C_2H_5{-}\overset{\displaystyle}{C}{-}CH_3 \xrightarrow{\text{heat}} C_2H_5\overset{\displaystyle}{C}HCH_3 + CO_2 \\
| \qquad\qquad\qquad | \\
\text{COOH} \qquad\qquad \text{COOH}
\end{array}
$$

Both nature and man make extensive use of this decarboxylation reaction. In synthetic work, it is possible to convert esters of malonic acid into substituted derivatives $R''OOC{-}CRR'{-}COOR''$. These esters can be hydrolysed to the corresponding dicarboxylic acids, which, when heated, provide monocarboxylic acids; this is a very useful method for making specific carboxylic acids. Nature uses such dicarboxylic acids in the same way; the enzyme-catalysed decarboxylation of substituted derivatives of malonic acid is an important process in the oxidation of biological compounds in metabolism to give carbon dioxide and water.

Succinic acid, $HOOCCH_2CH_2COOH$, and higher members of the series are not readily decarboxylated in this way.

Hydroxydicarboxylic acids also occur widely in nature, and play an important role in metabolism. Examples are malic acid,

$$\overset{\displaystyle OH}{\underset{\displaystyle |}{HOOCCH_2CHCOOH}}$$

which provides all of the acid content of apples, and tartaric acid, $HOOC-CHOH-CHOH-COOH$, whose potassium salt is present in grape juice and wines and is the chief constituent of the lees of wine. Citric acid, which occurs in lemons, blackcurrants, etc., is a hydroxytricarboxylic acid:

$$\begin{array}{c} CH_2COOH \\ | \\ HO-C-COOH \\ | \\ CH_2COOH \end{array}$$

The unsaturated acid fumaric acid also plays an important role in metabolism. It is the E (or *trans*) isomer; the Z (*cis*) isomer is known as maleic acid:

Fumaric acid   Maleic acid

Because of their different geometries these isomers are different in both physical properties, e.g. melting points, fumaric acid, $286\,°C$; maleic acid, $130\,°C$; and chemical properties (see next chapter). Furthermore, whereas maleic acid is rather toxic, fumaric acid is used as an acid in food products.

Questions

1. Write the structural formulae of (a) hexanoic acid, (b) 5-ethyl-3-methylheptanoic acid, (c) β-bromobutanoic acid, (d) α-hydroxypropionic acid and (e) hexadec-8-enoic acid.
2. Starting from bromomethane show how you could obtain (a) and (b) acetic acid, by two separate routes.
3. What reagents are used to bring about the following reactions?

   (a) $RCOOH \xrightarrow{?} RCH_2OH$

   (b) $RCOOH \xrightarrow{?} RCOCl$

   (c) $RCOOH \xrightarrow{?} RCO_2^- \overset{+}{N}H(C_2H_5)_3$

   (d) $RCOOH \xrightarrow{?} RCOOCH_3$

   Give a mechanism for reaction (d).

16 ACID CHLORIDES (ACYL CHLORIDES) AND ACID ANHYDRIDES

Acid Chlorides

The preparation and nomenclature of acid chlorides (or acyl chlorides), RCOCl, was described in the preceding chapter.

The simplest member of the series, formyl chloride HCOCl, is not stable at room temperature and decomposes to carbon monoxide and hydrogen chloride:

$$HCOCl \longrightarrow CO + HCl$$

It can be obtained at very low temperatures.

Acid chlorides are very sensitive to moisture and react violently with water. They fume in moist air. They are in general very reactive towards nucleophiles. In both the C=O bond (cf. aldehydes and ketones) and in the C—Cl bond (cf. alkyl halides) the electrons are drawn away from carbon and towards the oxygen and chlorine atoms respectively:

$$R-C \overset{\delta+}{\underset{\diagdown}{\diagup}} \overset{\delta-}{\overset{O}{\diagup}} Cl^{\delta-}$$

Hence a positive charge resides on this carbon atom and makes it very reactive towards nucleophiles. In addition, as in alkyl halides, Cl^- is a very good leaving group and is very readily displaced by the attacking nucleophile.

Thus the reaction with water, which converts an acid chloride into a carboxylic acid, is as follows:

$$CH_3C \overset{O}{\underset{Cl}{\rangle}} \quad :O \overset{H}{\underset{H}{\big\langle}} \quad \rightarrow \quad CH_3 - \overset{\overset{O^-}{|}}{\underset{\underset{Cl}{|}}{C}} - \overset{+}{O}H_2 \quad \rightarrow$$

$$CH_3 - \overset{\overset{O}{\|}}{C} - \overset{+}{O}H_2 \xrightarrow{-H^+} CH_3COOH$$

$$+ Cl^- \qquad\qquad \text{Acetic acid}$$

Hydroxide ions convert acid chlorides into the anion of the related carboxylic acid, for example

$$CH_3C \overset{O}{\underset{Cl}{\rangle}} \quad {}^-OH \atop Na^+ \quad \rightarrow \quad CH_3 - \overset{\overset{O^-}{|}}{\underset{\underset{Cl}{|}}{C}} - OH \quad \rightarrow$$

$$CH_3\overset{\overset{O}{\|}}{C} - OH \xrightarrow{HO^-} CH_3CO_2^- + H_2O$$

$$Na^+$$
$$\text{Sodium acetate}$$

Alcohols react similarly to water and provide esters

$$CH_3C \overset{O}{\underset{Cl}{\rangle}} \quad :O \overset{C_2H_5}{\underset{H}{\big\langle}} \quad \rightarrow \quad CH_3 - \overset{\overset{O^-}{|}}{\underset{\underset{Cl}{|}}{C}} - \overset{+}{O} \overset{C_2H_5}{\underset{H}{\big\langle}} \quad \rightarrow$$

$$CH_3 - \overset{\overset{O}{\|}}{C} - \overset{+}{O} \overset{C_2H_5}{\underset{H}{\big\langle}} \xrightarrow{-H^+} CH_3COOC_2H_5$$

Ammonia and amines are stronger nucleophiles than water and alcohols, and react violently with acid chlorides to give **amides**, for

example

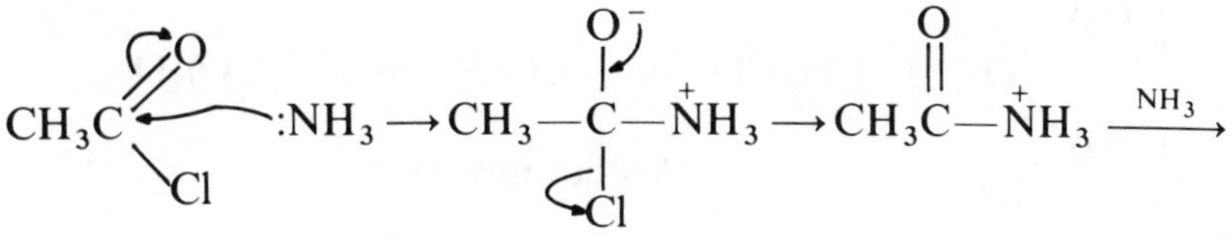

$$CH_3C \overset{O}{\underset{Cl}{\diagdown}} :NH_3 \rightarrow CH_3 - \overset{O^-}{\underset{Cl}{\overset{|}{C}}} - \overset{+}{NH_3} \rightarrow CH_3C \overset{O}{\overset{||}{-}} \overset{+}{NH_3} \xrightarrow{NH_3}$$

$$CH_3CONH_2 + \overset{+}{N}H_4$$

Acetamide

Amides are derivatives of carboxylic acids in which the hydroxy group has been replaced by an amino (NH_2) group or a substituted amino group (NHR or NR_2). Amides with substituted amino groups are formed when an amine reacts with an acid chloride, for example

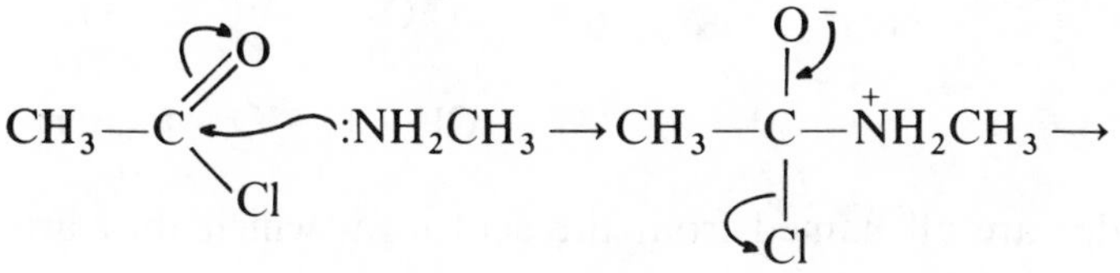

$$CH_3 - C \overset{O}{\underset{Cl}{\diagdown}} :NH_2CH_3 \rightarrow CH_3 - \overset{O^-}{\underset{Cl}{\overset{|}{C}}} - \overset{+}{N}H_2CH_3 \rightarrow$$

$$CH_3\overset{O}{\overset{||}{C}} - \overset{+}{N}H_2CH_3 \xrightarrow{NH_2CH_3} CH_3CONHCH_3 + \overset{+}{N}H_3CH_3$$

Another type of compound derived from carboxylic acids is called an **acid anhydride**:

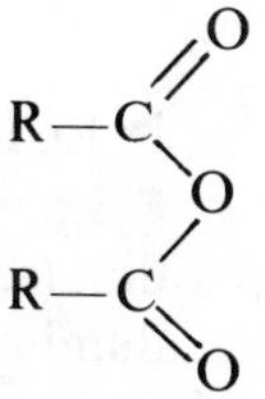

$$\begin{array}{c} R - C \overset{O}{\underset{O}{\diagup\diagdown}} \\ R - C \underset{O}{\overset{}{\diagdown}} \end{array}$$

An acid anhydride

The name is derived from the molecular formula, which adds up to two molecules of the corresponding carboxylic acid minus one molecule of water:

$$2\,RCOOH - H_2O = RCO - O - COR$$

The structural formula is often shortened to $(RCO)_2O$. An example is

$$CH_3C\overset{\displaystyle O}{\underset{\displaystyle O}{\diagup}}\diagdown O \diagup CH_3C\diagdown O$$

or $CH_3CO{-}O{-}COCH_3$ or $(CH_3CO)_2O$

Acetic anhydride

A compound of this structure could arise by replacement of the chlorine atom of acetyl chloride by an acetate group; reaction of sodium acetate with acetyl chloride does indeed provide acetic anhydride:

$$CH_3C\overset{O}{\diagup}\diagdown Cl \quad \overset{\delta-}{O} \overset{O}{\underset{\delta-}{\diagdown}} CCH_3 \quad Na^+ \quad \rightarrow CH_3{-}\overset{O^-}{\underset{Cl}{\overset{|}{C}}}{-}O{-}\overset{O}{\underset{||}{C}}CH_3$$

$$CH_3CO{-}O{-}COCH_3$$

Acid anhydrides are all named from the acid from which they are derived.

Another example of nucleophilic attack on acid chlorides is their reduction by lithium aluminium hydride or sodium borohydride:

$$RCOCl \overset{\text{LiAlH}_4}{\underset{\underset{\text{in pyridine}}{\text{NaBH}_4}}{\diagup \diagdown}} \overset{RCH_2OH}{\underset{RCHO}{}}$$

A solution of the borohydride in pyridine (see Chapter 28) reduces them only to an aldehyde, but lithium aluminium hydride reduces them further to a primary alcohol. Since both reagents in effect act as sources of hydride ions (see Chapter 14), a simplified mechanism for the reduction of an acid chloride may be expressed as

$$RC\overset{O}{\diagdown Cl}{\overset{-}{H}} \rightarrow R{-}\overset{O^-}{\underset{Cl}{\overset{|}{C}}}{-}H \rightarrow R\overset{O}{\underset{||}{C}}H$$

Acid Anhydrides

Acid anhydrides also react well with nucleophiles, but less violently than acid chlorides, as shown by the following examples involving acetic anhydride:

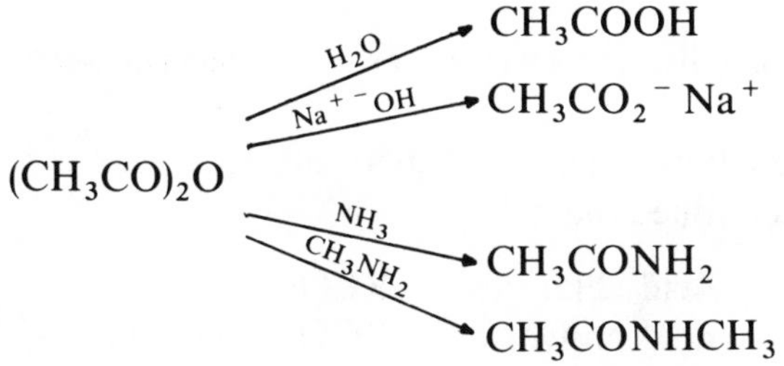

$$
(CH_3CO)_2O
\begin{cases}
\xrightarrow{H_2O} CH_3COOH \\
\xrightarrow{Na^+ \; {}^-OH} CH_3CO_2{}^- \; Na^+ \\
\xrightarrow{NH_3} CH_3CONH_2 \\
\xrightarrow{CH_3NH_2} CH_3CONHCH_3
\end{cases}
$$

It is not possible, however, to convert acid anhydrides into acid chlorides.

In the above reactions an acetate ion is displaced by the nucleophile and, since accetate ion is a good leaving group, the reaction proceeds readily. The mechanism is as exemplified by the reaction with water:

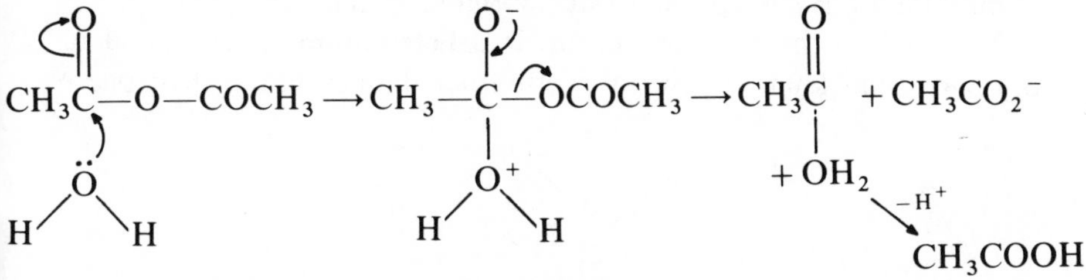

$$
CH_3\overset{O}{\overset{\|}{C}}-O-COCH_3 \rightarrow CH_3-\overset{O^-}{\overset{|}{C}}-OCOCH_3 \rightarrow CH_3\overset{O}{\overset{\|}{C}} + CH_3CO_2{}^-
$$

Some dicarboxylic acids having a suitable geometry to bring the two carboxyl groups near to one another are converted into cyclic anhydrides when they are heated. Examples are the formation of succinic anhydride and maleic anhydride.

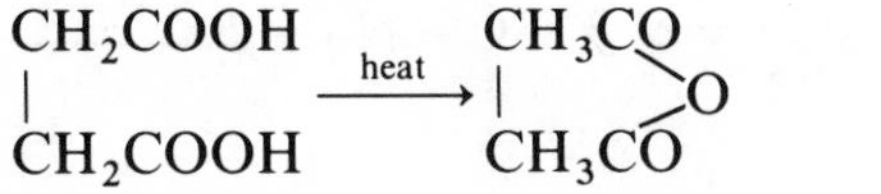

$$
\begin{array}{c}
CH_2COOH \\
| \\
CH_2COOH
\end{array}
\xrightarrow{heat}
\begin{array}{c}
CH_3CO \\
| \quad \diagdown O \\
CH_3CO
\end{array}
$$

Succinic acid   Succinic anhydride

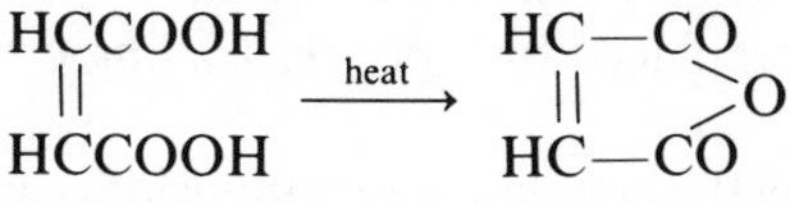

$$
\begin{array}{c}
HCCOOH \\
\| \\
HCCOOH
\end{array}
\xrightarrow{heat}
\begin{array}{c}
HC-CO \\
\| \quad \diagdown O \\
HC-CO
\end{array}
$$

Maleic acid   Maleic anhydride

Maleic acid, being a *Z* (*cis*) isomer, has the two carboxyl groups held near to each other. In the *E* (*trans*) isomer, fumaric acid, the two carboxyl groups are held far apart, and so a fumaric anhydride is not formed.

Comparison of Different Derivatives of Carboxylic acids

In this chapter four different kinds of derivatives of carboxylic acids have been mentioned, namely

Acid chlorides	RCOCl
Acid anhydrides	RCO—O—COR
Esters	RCOOR′
Amides	$RCONH_2$

All of these classes of compounds have structures containing a carbonyl group which, because of the unequal sharing of electrons in the $C{=}O$ bond, tends to have a partial positive charge on the carbon atom, making it susceptible to nucleophilic attack.

Attached to the carbonyl group is a **heteroatom**, a term used in organic chemistry to denote an atom other than carbon or hydrogen:

$$R-C{\overset{\displaystyle O}{\underset{\displaystyle Z}{<}}}$$

(Z = heteroatom)

A general equation for their reactions with nucleophiles is

$$R-C{\overset{\displaystyle O}{\underset{\displaystyle Z}{<}}} \underset{-Nu^-}{\overset{Nu^-}{\rightleftharpoons}} R-\overset{\displaystyle O^-}{\underset{\displaystyle Z}{\overset{|}{\underset{|}{C}}}}-Nu \underset{Z^-}{\overset{-Z^-}{\rightleftharpoons}} R-C{\overset{\displaystyle O}{\underset{\displaystyle Nu}{<}}}$$

(Nu^- = nucleophile)

If this is considered as two sets of equilibria it follows that a better leaving group can be replaced by a poorer leaving group, but not vice versa.

This is illustrated by the fact that acid anhydrides can be formed from acid chlorides but the latter cannot be converted into acid anhydrides, since chloride is a better leaving group than acetate and other carboxylate anions:

$$R-C\overset{\displaystyle O}{\underset{\displaystyle Cl}{<}} \quad + \quad {}^{-}O_2CR \quad \rightleftharpoons \quad R-C\overset{\displaystyle O}{\underset{\displaystyle OCOR}{<}} \quad + \, Cl^{-}$$

If we consider acid chlorides (Z=Cl), acid anhydrides (Z=RCO$_2$), esters (Z=OR') and amides (Z=NH$_2$), then the character of these groups (Z) as leaving groups diminishes in the order $Cl^- > RCO_2^- > R'O^- > H_2N^-$.

In other words, as we have seen, acid chlorides can be converted into anhydrides, esters or amides, whereas acid anhydrides can be converted into esters or amides, but **not** into acid chlorides. We may similarly expect that esters can be converted only into amides and that amides cannot be converted into any of these other derivatives.

Their reactivity towards nucleophiles, e.g. to hydrolysis by water, diminishes in the same order, namely $RCOCl > RCO-O-COR > RCOOR' > RCONH_2$.

Another factor contributing to these differences in reactivity is the ability of the lone pair of electrons in $RCO\ddot{Z}$ to lower the positive charge on the carbon atom by interaction with it. An extreme case of this was seen for the carboxylate anion:

$$R-C\overset{\displaystyle O}{\underset{\displaystyle O^-}{<}} \quad \longleftrightarrow \quad R-C\overset{\displaystyle O^-}{\underset{\displaystyle O}{<}}$$

In neutral molecules the ability of a lone pair of electrons on an atom $\ddot{Z}$ to interact with a positively charged centre decreases in the order $N > O > Cl$. In the case of the acyl derivatives this means that this effect is seen most in amides and least in acid chlorides. Thus there will be more neutralization of the positive charge on the carbonyl carbon atom in the case of amides than in acid chlorides, resulting in a further contribution to the decreased reactivity towards nucleophiles of amides as compared to acid chlorides.

The effect of the oxygen atom in acid anhydrides and esters will be intermediate between that of the nitrogen and chlorine atoms. The effect in an ester should be more marked than in an anhydride since in the former there is one oxygen atom joined to each carbonyl group, whereas in an anhydride there is only one oxygen atom joined to two carbonyl groups:

$$R-C\underset{OR'}{\overset{O}{\diagup}}\qquad \begin{array}{l}\text{Ester}\\ 1\quad C{=}O,\ 1\quad -O-\end{array}$$

$$R-C\overset{O}{\diagup}\underset{O}{\diagdown}C-R\qquad \begin{array}{l}\text{Anhydride}\\ 2\quad C{=}O,\ 1\quad -O-\end{array}$$

Hence any neutralization of the positive charge on the carbonyl group should be more effective in an ester than in an anhydride.

The overall effect of this electron donation from a neighbouring atom on the reactivity of the carbonyl group works in the same direction as does the effectiveness as a leaving group, reinforcing the order of reactivity towards nucleophiles $RCOCl > (RCO)_2O > RCOOR' > RCONH_2$.

This difference in reactivity may be compared with the different reactivities of alkyl halides, alcohols and ethers, and amines towards nucleophiles, as discussed in Chapter 11.

Some of the chemistry of esters and amides is discussed in succeeding chapters.

Questions

1. How may propionyl chloride be converted into (a) ethyl propionate, (b) propionic anhydride and (c) propionaldehyde? Give the mechanisms of the reactions.
2. Starting from acetic anhydride how could you obtain (a) acetic acid, (b) potassium acetate and (c) N-methylacetamide?

17 ESTERS

Esters can be prepared by the reactions of alcohols with either carboxylic acids in the presence of an acid catalyst, acid chlorides or acid anhydrides, as discussed in Chapters 15 and 16, or by the reaction between an alkyl halide and the salt of a carboxylic acid, as mentioned in Chapter 7.

Nomenclature

Esters are named simply by using the name of the alkyl group that takes the place of the acidic hydrogen of the acid and the name of the anion derived from the acid, for example

$$H-C\overset{\displaystyle O}{\underset{\displaystyle OC_2H_5}{\Big\backslash}} \qquad CH_3(CH_2)_4C\overset{\displaystyle O}{\underset{\displaystyle OCH_3}{\Big\backslash}} \qquad \begin{array}{c} CH_3COOCH_2 \\ | \\ CH_3COOCH \\ | \\ CH_3COOCH_2 \end{array}$$

Ethyl formate   Methyl hexanoate   Glyceryl triacetate

Although the name includes the name of the anion of the acid, it must of course be remembered that no such ion is actually present; the whole ester molecule is covalently linked.

Isomerism

Isomerism is obviously possible. For example methyl propionate, $CH_3CH_2COOCH_3$, is isomeric with ethyl acetate, $CH_3COOCH_2CH_3$.

In this case all that has been done is to exchange the two alkyl groups R and R' in the general formula RCOOR'. However, these are not the only isomers with this molecular formula $C_4H_8O_2$. Two other esters have this formula, $HCOOCH_2CH_2CH_3$, propyl formate, and $HCOOCHCH_3$, isopropyl formate. [Isomeric carboxylic acids,
|
CH_3
$CH_3CH_2CH_2COOH$ and $(CH_3)_2CHCOOH$, also have the same molecular formula.]

Distinction between an ester and a carboxylic acid is very simple because of the acidity of the latter but not of the ester. The simplest way to distinguish between different esters is commonly from their n.m.r. spectra. For example, although methyl propionate and ethyl acetate each contains a methyl group and an ethyl group, these groups are in different chemical environments in the two esters. The methyl group is directly attached to an oxygen atom in the former and to a carbon atom in the latter, and these signals will hence appear at different positions, which can be checked from literature tables of chemical shifts.

Reactions of Esters

Esters are attacked by good nucleophiles, as discussed in the previous chapter:

$$R-C\overset{O}{\underset{OR'}{\diagdown}} \quad Nu^- \;\longrightarrow\; R-\overset{O^-}{\underset{OR'}{\overset{|}{C}}}-Nu \;\longrightarrow\; R-C\overset{O}{\underset{Nu}{\diagdown}} \;+\; {}^-OR'$$

Water attacks esters and **hydrolyses** them to form carboxylic acids and alcohols, for example

$$CH_3C\overset{O}{\underset{OC_2H_5}{\diagdown}} \quad\overset{H_2O}{\longrightarrow}\quad CH_3C\overset{O}{\underset{OH}{\diagdown}} \;+\; C_2H_5OH$$

Hydrolysis is a reaction in which a molecule is cleaved by reaction with water into two or more parts. A hydrogen atom from the water adds to one of the fragments and a hydroxy group to the other.

The mechanism of the hydrolysis of esters is

$$CH_3C\overset{O}{\underset{OC_2H_5}{\diagup}} \ + \ :O\overset{H}{\underset{H}{\diagdown}} \rightleftharpoons CH_3\overset{O^-}{\underset{OC_2H_5}{\overset{|}{C}}}-\overset{+}{O}\overset{H}{\underset{H}{\diagdown}} \rightleftharpoons CH_3\overset{O}{\overset{||}{C}}-OH$$

$$\overset{\diagdown}{O^+}\overset{H}{\diagup}\overset{}{\diagdown}C_2H_5$$

$$CH_3COOH + C_2H_5OH$$

This is the exact reverse of the mode of formation of an ester from an alcohol and an acid, and is controlled by the same equilibrium constant:

$$K = \frac{[\text{ester}]\,[H_2O]}{[\text{alcohol}]\,[\text{acid}]}$$

Hydrolysis of the ester is favoured by having an excess of water present. To maintain K as a constant, this necessitates that the quantity of [ester] decreases and the quantities of [alcohol] and [acid] increase.

As with ester formation, ester hydrolysis is catalysed by the presence of acid. The carbonyl group is protonated and attack by water is thereby facilitated:

$$CH_3C\overset{O}{\underset{OC_2H_5}{\diagup}} \ \underset{}{\overset{H^+}{\rightleftharpoons}} \ CH_3\overset{+}{\underset{OC_2H_5}{C}}\overset{OH}{\diagup} \ + \ :O\overset{H}{\underset{H}{\diagdown}} \rightleftharpoons CH_3\overset{OH}{\underset{OC_2H_5}{\overset{|}{C}}}-\overset{+}{O}\overset{H}{\underset{H}{\diagdown}}$$

$$CH_3C\overset{O}{\underset{OH}{\diagup}} \ \underset{}{\overset{H_2O}{\rightleftharpoons}} \ CH_3\overset{OH}{\underset{+}{\overset{|}{C}}}-O-H \ \rightleftharpoons \ CH_3\overset{OH}{\underset{}{\overset{|}{C}}}-OH$$

$$+ H_3O^+ \qquad\qquad + C_2H_5OH \qquad\qquad \overset{\diagdown}{O^+}\overset{H}{\diagup}\overset{}{\diagdown}C_2H_5$$

Esters can also be hydrolysed in basic conditions. This reaction is often called the **saponification** of an ester, since it is the process that has been used for centuries in the manufacture of soap (described later in this chapter):

$$CH_3C\underset{OC_2H_5}{\overset{O}{\diagup}} \quad \overset{-}{O}H \longrightarrow CH_3\overset{\overset{O^-}{|}}{\underset{OC_2H_5}{C}}—OH \qquad CH_3C\underset{OH}{\overset{O}{\diagup}} \qquad CH_3CO_2^-$$

$$+ \qquad \longrightarrow \qquad +$$

$$^-OC_2H_5 \qquad HOC_2H_5$$

This reaction is irreversible and does not involve an overall equilibrium because in the presence of base no reaction takes place between the products, an alcohol and a salt.

Furthermore, in saponification of an ester, it is frequently easy to separate the products. Salts are involatile, most simple alcohols are volatile, and hence they can be separated. Acidification of a solution of the salt provides the carboxylic acid.

Before the introduction of spectroscopic techniques esters were commonly identified by saponifying them and identifying the acid and alcohol from which they were derived.

An RO^- group is a better leaving group than NH_2^-, so esters react with ammonia to give amides, for example

$$CH_3C\underset{OC_2H_5}{\overset{O}{\diagup}} :NH_3 \rightarrow CH_3\overset{\overset{O^-}{|}}{\underset{OC_2H_5}{C}}—\overset{+}{N}H_3 \rightarrow CH_3\overset{\overset{O}{\parallel}}{C}—\overset{+}{N}H_3 \rightarrow CH_3\overset{\overset{O}{\parallel}}{C}—NH_2$$

$$+ \; ^-OC_2H_5 \quad HOC_2H_5$$

(This is probably a simplified version of the precise mechanism.)

Esters (and also acid anhydrides and acid chlorides) react with Grignard reagents to form ketones, which are transformed further into tertiary alcohols:

$$CH_3C\underset{OC_2H_5}{\overset{O}{\diagup}} + RMgX \rightarrow CH_3\overset{\overset{OMgX}{|}}{\underset{OC_2H_5}{C}}—R \xrightarrow[H_2O]{H^+} CH_3\overset{\overset{O—H}{|}}{\underset{R}{C}}—OC_2H_5$$

$$CH_3\overset{\overset{OH}{|}}{\underset{R}{C}}—R \xleftarrow[(2)H^+,H_2O]{(1)\,RMgX} CH_3C\underset{R}{\overset{O}{\diagup}} + HOC_2H_5$$

Lithium aluminium hydride reduces esters to two molecules of alcohols, one derived from the alkyl part and one from the acyl part of the esters:

$$CH_3COOCH_3 \xrightarrow{\text{LiAlH}_4} CH_3CH_2OH + CH_3OH$$

$$C_2H_5COOCH_3 \xrightarrow{\text{LiAlH}_4} C_2H_5CH_2OH + CH_3OH$$

$$CH_3COOC_2H_5 \xrightarrow{\text{LiAlH}_4} 2C_2H_5OH$$

Physical Properties of Esters

Esters have no hydroxy group and thus cannot form hydrogen bonds in the way that alcohols or carboxylic acids do. For this reason most esters are insoluble in water. The lowest members of the series, methyl formate and methyl acetate, dissolve, because of interaction with water molecules:

$$\underset{CH_3\overset{\overset{\displaystyle O}{\|}}{C}OCH_3}{\cdots\cdots HOH}$$

However, as the size of the alkyl groups increases, solubility in water disappears. Ethyl acetate is only slightly soluble.

Most esters of low or medium molecular weight are liquids at room temperature.

Uses of Esters

Esters are widely distributed in nature. Many simpler esters have fragrant smells and make large contributions to natural perfumes and flavours. Synthetic esters also find large use as artificial perfumes and flavourings.

Plant and animal **waxes** contain esters of high molecular weight, RCOOR', in which both R and R' are made up from long unbranched chains of carbon atoms. These chains are frequently 20–30 carbon atoms in length, or even longer, and the acid and alcohol components usually each have even numbers of carbon atoms. The chains may contain double bonds. Common examples are beeswax and wool wax.

Fats and Oils

Animal and plant fats and oils also consist to a large extent of esters, but commonly they are esters of the trihydric alcohol glycerol and are hence called **glycerides.** They have the general formula

$$RCOOCH_2$$
$$|$$
$$R'COOCH$$
$$|$$
$$R''COOCH_2$$

R, R' and R" may or may not be the same. They may be saturated chains of carbon atoms or they may contain double bonds. Examples of common acid components are

$$CH_3(CH_2)_{14}COOH \qquad \text{Palmitic acid}$$
$$CH_3(CH_2)_{16}COOH \qquad \text{Stearic acid}$$

$$\underset{CH_3(CH_2)_7}{\overset{H}{\diagdown}}C=C\underset{(CH_2)_7COOH}{\overset{H}{\diagup}} \qquad \text{Oleic acid}$$

$$\underset{CH_3(CH_2)_4}{\overset{H}{\diagdown}}C=C\underset{CH_2}{\overset{H}{\diagup}}\underset{}{\overset{H}{\diagdown}}C=C\underset{(CH_2)_7COOH}{\overset{H}{\diagup}} \qquad \text{Linoleic acid}$$

In the body, fats and oils are both converted into carbon dioxide and water, providing energy in the process, and are used for building tissues, cell walls, etc. They thus have an enormous use and importance as foods.

In general land animals' fats have fewer double bonds in their structure than do the fats from marine animals or vegetable oils.

Also, in general terms, an increase in the number of double bonds in fats results in their having lower melting points. This is also the case for the constituent acids, e.g. stearic, oleic and linoleic acids, all of which contain the same number of carbon atoms and have melting points, respectively, 70°C, 13°C and − 5 °C. Partly as a consequence of these two facts, beef fat is harder than whale oil, olive oil or sunflower oil.

It also follows that oils and fats can be hardened by lowering the number of double bonds they contain, i.e. by hydrogenating them chemically. Hydrogenation of oils, over suitable catalysts, to make them harder is a large-scale industrial process, and in this way margarine and cooking fats are produced from oils.

Another important use of fats and oils is in soap-making, in which they are **saponified** by treatment with sodium hydroxide. The resultant sodium salts of the long-chain carboxylic acids are soaps:

$$
\begin{array}{ccc}
\text{RCOOCH}_2 & \text{RCO}_2^-\ \text{Na}^+ & \text{CH}_2\text{OH} \\
| & & | \\
\text{R'COOCH} \xrightarrow{\ \text{NaOH}\ } & \text{R'CO}_2^-\ \text{Na}^+ + & \text{CHOH} \\
| & & | \\
\text{R''COOCH}_2 & \text{R''CO}_2^-\ \text{Na}^+ & \text{CH}_2\text{OH} \\
 & \text{Soaps} & \text{Glycerol}
\end{array}
$$

The crude soap is skimmed and/or filtered off, purified, and perfume is added for domestic use.

A very crude explanation of the efficacy of soap as a cleansing agent is as follows. Most of the soap molecule, namely its carbon chain, is soluble in organic materials but not in water, whereas the carboxylate end of the molecule can associate with water by hydrogen bonding. If there is a layer of greasy dirt on a surface it does not dissolve in water applied to it. If soap is present in the water, the soap molecules arrange themselves so that the carbon chain associates with the grease while the carboxylate group remains in the water. This can be represented diagrammatically thus:

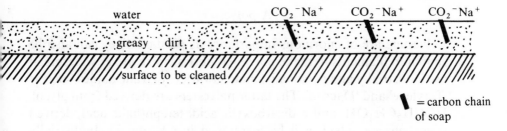

= carbon chain
of soap

If the water layer is rinsed off (or wiped off with a towel!) it takes the carboxylate groups with it. They in turn take their own carbon chains along, and the dirt adheres to these chains. Thus the dirt is removed from the surface to be cleaned and is carried away with the water.

Many synthetic detergents are salts of **sulphonic acids**. RSO_3^- Na^+. They act in the same way, with the molecules orienting themselves so that the alkyl chain R is in the grease and the **sulphonate** group, $—SO_3^-$, is the water. Soap may cause scum in hard water areas because calcium and magnesium salts of soap which are formed are insoluble in water. Here the synthetic detergents may be advantageous, in that their calcium and magnesium salts can be soluble in water and hence do not form scums.

As mentioned in Chapter 9, glycerol, the other product from soap manufacture, has a large use as a moistening agent, and for the manufacture of plastics and of the explosive glyceryl trinitrate, commonly known as nitroglycerine.

Polyesters

In the case of glycerol, each hydroxy group can react with a molecule of acid, i.e. glycerol forms triesters.

Supposing one considers the reaction of a diol, having two alcohol groups, with a dicarboxylic acid. Let us draw the reactants as HO ⁓ OH and HOOC ⁓ COOH, where the wavy line represents in a general way the carbon atoms between the two groups. These reactants could interact as follows to give a condensation polymer, which is a **polyester**:

etc. + HO ⁓ OH + HOOC ⁓ COOH + HO ⁓ OH + HOOC ⁓ COOH etc.

$$\downarrow$$

⁓ COO ⁓ OOC ⁓ COO ⁓ OOC ⁓ COO ⁓

A polyester

Polyesters are of enormous importance as plastics. They can also be spun into filaments to make polyester fabrics which are in widespread use industrially and personally, under a variety of trade names, such as 'Terylene' and 'Dacron'. The latter polyesters are derived from glycol, $HOCH_2CH_2OH$, and a dicarboxylic acid, terephthalic acid, derived from benzene, which will be mentioned in Chapter 27. Industrially these polymers are often made not directly from the diol and diacid; a diester is often used instead of the diacid to react with the diol.

The role played by polyesters in everyday life is enormous. For example, very few of us do not wear at least some clothing made from fabrics containing polyesters.

Questions

1. The reactions between alcohols and acids which lead to the formation of esters are equilibria. Write the equation for the equilibrium constant that governs the reaction between acetic acid and ethanol.
2. What conditions would you use to effect (a) the maximum conversion of a carboxylic acid into an ester and (b) the conversion of an ester into a carboxylic acid?
3. Compound **A** contains 58.8% carbon, 9.8% hydrogen and 31.4% oxygen, and has a molecular weight of 102. It shows four signals in its ^{13}C-n.m.r. spectrum and three signals in its ^{1}H-n.m.r. spectrum. With dilute aqueous acid compound **A** is converted into compounds **B** and **C**.

   Compound **B** shows two signals in its ^{13}C-n.m.r. spectrum and three signals in its ^{1}H-n.m.r. spectrum, the ratio of the intensities of the latter being 6:1:1. Compound **C** shows two signals in its ^{13}C-n.m.r. spectrum and two signals (intensities 3:1) in its ^{1}H-n.m.r. spectrum. What are the structural formulae of **A**, **B** and **C**? Give the mechanism of the reaction

$$(A) \xrightarrow[\text{aqueous acid}]{\text{dilute}} (B) + (C)$$

18 ESTERS OF INORGANIC ACIDS

The esters discussed in the previous chapter were all derived from alcohols and organic carboxylic acids. Esters can also be made from alcohols and inorganic acids, and a few examples of inorganic esters are mentioned in this chapter.

In a sense alkyl halides are esters of the inorganic halogen acids. Thus

$$C_2H_5OH + HI \longrightarrow C_2H_5I + H_2O$$

However, the mechanism and mode of formation of the alkyl halide is different from that which is involved in the formation of esters as described earlier, and in any case, it is convenient and practical to consider alkyl halides separately.

The esters considered here are derived from inorganic oxyacids, i.e. inorganic acids that ionize to give protons and anions in which negative charge resides on an oxygen atom:

$$X-O-H \rightleftharpoons X-O^- + H^+.$$

Examples are sulphuric acid, $SO_2(OH)_2$, nitric acid, $HONO_2$, nitrous acid, $HONO$, and phosphoric acid, $(HO)_3PO$.

As in the case of carboxylic acids, a hydrogen atom of the acid is replaced by an alkyl group to give an ester, as in $C_2H_5ONO_2$, ethyl nitrate.

Esters of Sulphuric Acid

Reaction of sulphuric acid with alcohols can give monoesters or diesters, for example

$$SO_2(OH)_2 + CH_3OH \longrightarrow \underset{\underset{OCH_3}{|}}{\overset{\overset{OH}{|}}{SO_2}} \xrightarrow{CH_3OH} \underset{\underset{OCH_3}{|}}{\overset{\overset{OCH_3}{|}}{SO_2}}$$

Methyl hydrogen   Dimethyl
sulphate or   sulphate
methyl sulphuric acid

It is important that the reaction is kept cold or alternative reactions may take place (see Chapter 9).

Since the monoesters retain one of the acidic hydrogen atoms, they are themselves strong acids.

Monoesters made from long-chain alcohols are converted into sodium salts, which are of commercial importance as detergents:

$$ROSO_2OH + Na_2CO_3 \longrightarrow ROSO_2O^- \quad Na^+$$

The negative charge in the anion is in fact delocalized over the three oxygen atoms which are not involved in bonding to the alkyl group, and the anion is often written as $ROSO_3^-$.

These monoesters can also be made from alkenes, as described in Chapter 5, for example

$$CH_2{=}CH_2 \xrightarrow{H_2SO_4} CH_3CH_2OSO_2OH$$

Monoesters and diesters of sulphuric acid react very readily with nucleophiles, but the reaction is like that of an alkyl halide rather than that of an organic ester:

$$Nu^- \overset{\frown}{\quad} R \overset{\frown}{\quad} OSO_2OH \longrightarrow NuR + {}^-OSO_2OH$$
$$(Nu^- = nucleophile)$$

As a specific example we can have

$$Na^+ \quad HO^- \overset{\frown}{\quad} CH_3 \overset{\frown}{\quad} OSO_2OH \longrightarrow HOCH_3 + {}^-OSO_2OH \quad Na^+$$
Sodium hydrogen
sulphate

Mechanistically alkyl sulphates react with nucleophiles in the same way as alkyl halides do and, since they are very reactive, they are sometimes used instead of alkyl halides:

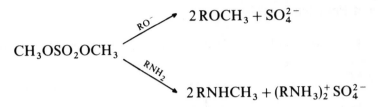

$$CH_3OSO_2OCH_3 \xrightarrow{RO^-} 2\,ROCH_3 + SO_4^{2-}$$

$$\xrightarrow{RNH_2} 2\,RNHCH_3 + (RNH_3)_2^+ SO_4^{2-}$$

All esters derived from strong inorganic acids behave similarly. Since the leaving groups are anions derived from strong acids they are particularly good leaving groups.

One consequence of the high reactivity of sulphate esters towards nucleophiles is that they can be dangerously toxic materials, because they react with hydroxy groups in the body. Dimethyl sulphate has had considerable use as a **methylating agent** for converting hydroxy groups into methyl ethers, especially in the study of sugars and carbohydrates:

$$ROH \xrightarrow{(CH_3)_2SO_4} ROCH_3$$

Great care must be taken in using dimethyl sulphate. It is volatile and its vapours are toxic and in addition it can be absorbed through the skin.

Esters of Nitric Acid

The reaction of glycerol with nitric acid to form the explosive glyceryl trinitrate has already been mentioned. All alcohols can react with nitric acid to form esters, but these reactions can be extremely dangerous and result in violent explosions. One difficulty is that nitric acid can also act as an oxidizing agent, and this reaction can proceed with great violence.

Esters of Nitrous Acid

Some esters of nitrous acid are used medically to provide relief to sufferers from *angina pectoris*. They relax the smooth muscles of the body and produce a lowering of blood pressure. They may be made by reaction of an alcohol with an inorganic nitrite salt in the presence of a mineral acid, for example

$$C_2H_5OH + NaNO_2 + HCl(\text{or } H_2SO_4) \longrightarrow C_2H_5ONO$$
$$\text{Ethyl nitrite}$$

Esters of Phosphoric Acid

Since phosphoric acid is a tribasic acid it can form mono-, di- and triesters for example

$$C_2H_5O-\overset{\overset{\displaystyle OH}{|}}{\underset{\underset{\displaystyle O}{\|}}{P}}-OH \qquad C_2H_5O-\overset{\overset{\displaystyle OH}{|}}{\underset{\underset{\displaystyle O}{\|}}{P}}-OC_2H_5 \qquad C_2H_5O-\overset{\overset{\displaystyle OC_2H_5}{|}}{\underset{\underset{\displaystyle O}{\|}}{P}}-OC_2H_5$$

Ethyl phosphate   Diethyl phosphate   Triethyl phosphate

Since the mono- and diesters still contain hydroxy groups they are both acidic. Triesters are usually made from phosphorus oxychloride:

$$POCl_3 + 3\,ROH \longrightarrow PO(OR)_3$$

When phosphoric acid is heated it is converted into pyrophosphoric acid, which is, in effect, the anhydride of phosphoric acid:

$$HO-\overset{\overset{\displaystyle O}{\|}}{\underset{\underset{\displaystyle OH}{|}}{P}}-OH \xrightarrow{\text{heat}} HO-\overset{\overset{\displaystyle O}{\|}}{\underset{\underset{\displaystyle OH}{|}}{P}}-O-\overset{\overset{\displaystyle O}{\|}}{\underset{\underset{\displaystyle OH}{|}}{P}}-OH$$

Pyrophosphoric acid

Pyrophosphoric acid still contains hydroxy groups and thus also forms esters with alcohols.

   Esters of phosphoric acid and of pyrophosphoric acid are extremely important in nature. Since both of these acids are strong acids, their anions are very good leaving groups and nature makes great use of them as such in biological chemical reactions. They play the same role in nature as the alkyl halides play in laboratory chemistry, reacting efficiently with nucleophiles in both substitution and elimination reactions.

Esters of Carbonic Acid

Esters of carbonic acid are made from the acid chloride of carbonic acid, $COCl_2$, which is **carbonyl chloride**, perhaps best known as the poisonous gas **phosgene**, used extensively in the 1914–1918 World War:

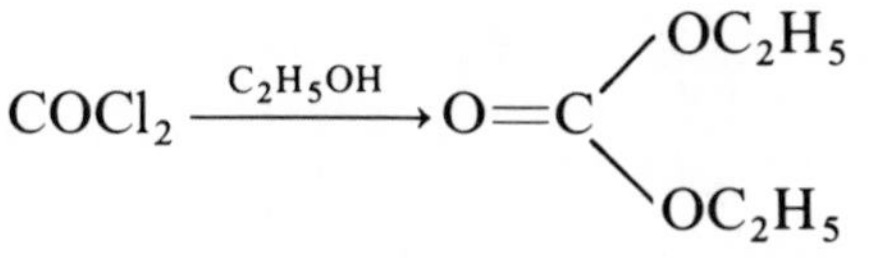

$$COCl_2 \xrightarrow{C_2H_5OH} O{=}C\Big\langle \begin{matrix} OC_2H_5 \\ OC_2H_5 \end{matrix}$$

Diethyl carbonate

Of great importance as plastics are **polycarbonates**. Since carbonic acid is a dibasic acid it can react with diols to give these polyesters, although they are not in fact made by this direct method industrially:

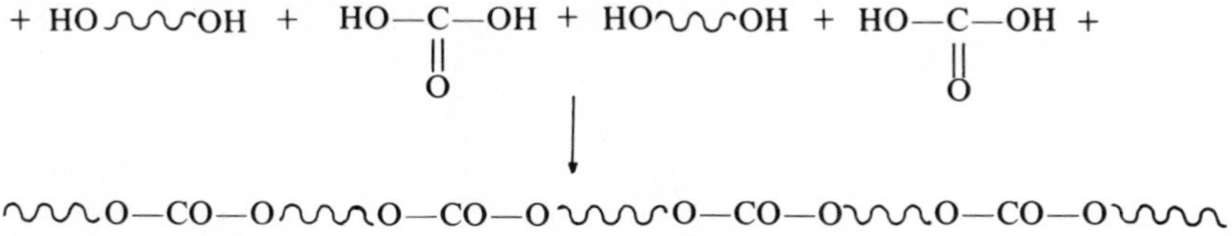

$$+ \text{HO}\!\sim\!\!\sim\!\!\text{OH} + \text{HO}{-}\underset{\underset{O}{\|}}{C}{-}\text{OH} + \text{HO}\!\sim\!\!\sim\!\text{OH} + \text{HO}{-}\underset{\underset{O}{\|}}{C}{-}\text{OH} +$$

$$\downarrow$$

$$\sim\!\!\sim\!\!O{-}CO{-}O\!\sim\!\!\sim\!O{-}CO{-}O\sim\!\!\sim\!\!O{-}CO{-}O\sim\!\!\sim\!O{-}CO{-}O\sim\!\!\sim$$

A 'polycarbonate'

Polycarbonates are used extensively as substitutes for glass in bottles, vandal-proof 'glass' sheets, helmets and shields. Some epoxy resins, used as adhesives, are also polycarbonates.

19 AMIDES

Amides have the general formula $RC\overset{\diagup O}{\underset{\diagdown NH_2}{}}$, or $RCONR'_2$ for N-substituted amides. They are prepared by reaction of acid chlorides, anhydrides or esters with ammonia or amines:

$$\left.\begin{array}{c} RCOCl \\ or \\ (RCO)_2O \\ or \\ RCOOR' \end{array}\right\} \begin{array}{l} \xrightarrow{NH_3} RCONH_2 \\[1ex] \xrightarrow{CH_3NH_2} RCONHCH_3 \\[1ex] \xrightarrow{(CH_3)_2NH} RCON(CH_3)_2 \end{array}$$

Amides are also formed when ammonium salts of carboxylic acids are heated, for example

$$CH_3CO_2^- \ \overset{+}{N}H_4 \xrightarrow{heat} CH_3CONH_2$$

Ammonium acetate   Acetamide

Amides are named after the acid from which they derive, replacing '-ic acid' or '-oic acid' by '-amide', for example

CH_3CONH_2	Acetamide
$CH_3(CH_2)_5CONH_2$	Hexanamide
$HCON(CH_3)_2$	N,N-dimethylformamide

In the latter name the initial N,N- shows that each of the methyl groups take the place of a hydrogen atom at the nitrogen atom. N,N-dimethylformamide is widely used as a solvent.

Lower members of the amide series are readily soluble in water; like carboxylic acids they are strongly hydrogen bonded by water.

Unlike amines, amides are not basic, but neutral. The lone pair of electrons associated with the nitrogen atom, which is responsible for the basicity of amines, in the case of amides interacts instead with the adjacent carbonyl group:

$$R-C\underset{\overset{..}{NH_2}}{\overset{O}{\diagup}} \quad \longleftrightarrow \quad R-C\underset{\underset{+}{NH_2}}{\overset{O^-}{\diagup}}$$

or, in dotted line nomenclature:

$$R-C\underset{\underset{\delta+}{NH_2}}{\overset{O^{\delta-}}{\diagup}}$$

This interaction results in there being some double-bond character in the C—N link. This can be seen from the n.m.r. spectra of N-substituted amides. For example, in the ^{1}H- n.m.r. or ^{13}C-n.m.r. spectra of N,N-dimethylformamide, two signals corresponding to methyl groups are observed:

$$R-C\overset{O^{\delta-}}{\underset{\underset{\delta+}{N}}{\diagup}}\diagup\overset{CH_3}{\underset{CH_3}{}}$$

Because of the partial double-bond character in C⚌N, rotation about this bond, which would be very fast if it were a simple single bond, is restricted, and so one methyl group is pointing towards the carbonyl group and one points away from it. Consequently they are in different chemical environments and give two separate signals.

Like water, amides are both neutral and amphoteric. In the absence of water they can form cations and anions:

If any water is present these ions are reconverted into the amide.

As has been seen in Chapter 16, amides are the least reactive of the common carboxylic acid derivatives towards nucleophiles. The only reactions considered here are their hydrolysis and reduction.

Amides can be hydrolysed to carboxylic acids, but not easily, and either base or acid catalysis is required. These catalysed reactions can be summarized as follows:

In each case the important step is the loss of NH_3, which is a good leaving group.

Lithium aluminium hydride, acting as a source of hydride ions, reduces amides to amines:

$$RCONH_2 \xrightarrow{\text{LiAlH}_4} RCH_2NH_2$$

Amides of Inorganic Acids

The previous chapter discussed esters of inorganic acids; these acids can also give rise to amides. As with carboxylic acids, this involves replacement of a hydroxy group in the acid by an amino (NH_2, NHR or NR_2) group. Thus an amide from sulphuric acid would have a formula $HOSO_2NHR$ (or $HOSO_2NH_2$ or $HOSO_2NR_2$). Sulphonic acids (see Chapter 25), which have the formula RSO_2OH, provide **sulphon-amides**, which are very effective anti-bacterial agents. They are made from the corresponding acid chloride:

$$RSO_2Cl + NHR'_2 \longrightarrow RSO_2NR'_2$$

Nitric acid, $HONO_2$, forms amides, $RNHNO_2$, which are also called **nitramines**. Some of these are used as explosives, in particular

which is known as RDX and is one of the most important military explosives. Nitramines are commonly made by the action of nitric acid on amines:

$$RNH_2 + HONO_2 \longrightarrow RNHNO_2$$

Amines can react similarly with nitrous acid, or a nitrite salt in the presence of acid, to give **nitrosamines**. This is a reaction of biological significance, since many nitrosamines can act as carcinogens.

The amide of carbonic acid, $CO(NH_2)_2$, is a compound of importance, called **urea**. It is made industrially from carbon dioxide and ammonia, and has a variety of uses. It is itself used on a large scale as a fertilizer and source of nitrogen for plants. It is also used for making condensation polymers, especially from reactions with formaldehyde. Plastics so obtained are used for making, among other things, glues, bonded woods and domestic products such as formica and tableware. It is also used for making barbiturates used as sedatives, by reactions with substituted esters of malonic acid. An example is

$$
\begin{array}{ccc}
C_2H_5 \diagdown \quad \diagup COOC_2H_5 & NH_2 & C_2H_5 \diagdown \quad \diagup CO-NH \\
\quad\quad C & + \quad C=O \rightarrow & \quad\quad C \qquad CO \\
C_2H_5 \diagup \quad \diagdown COOC_2H_5 & NH_2 & C_2H_5 \diagup \quad \diagdown CO-NH
\end{array}
$$

Diethyl α, α-diethylmalonate   Urea   A barbiturate

Polyamides

In Chapter 17 it was seen that dicarboxylic acids and diols could undergo condensation to form polyesters. Similarly there are polyamides. If a mixture of a diacid and a diamine is heated, the salt that is formed decomposes to give a polyamide:

$$H_2N \rightsquigarrow NH_2 + HOOC \rightsquigarrow COOH + H_2N \rightsquigarrow NH_2 + HOOC \rightsquigarrow COOH$$

$$\downarrow$$

$$H_3\overset{+}{N} \rightsquigarrow \overset{+}{N}H_3 \quad {}^-O_2C \rightsquigarrow CO_2^- \quad H_3\overset{+}{N} \rightsquigarrow \overset{+}{N}H_3 \quad {}^-O_2C \rightsquigarrow CO_2^-$$

$$\downarrow \text{heat}$$

$$\rightsquigarrow CO-NH \rightsquigarrow NH-CO \rightsquigarrow CO-NH \rightsquigarrow NH-CO \rightsquigarrow CO-NH \rightsquigarrow$$

Nylons are polyamides and are of enormous commercial importance. Common starting materials for their preparation are 1,6-

diaminohexane, $H_2N(CH_2)_6NH_2$,   and   adipic   acid, $HOOC(CH_2)_4COOH$. They react to form a nylon of structure

$$\sim\!\sim NHCO(CH_2)_4CONH(CH_2)_6NHCO(CH_2)_4CONH(CH_2)_6NHCO \sim\!\sim$$

Because each reactant has six carbon atoms this is known as Nylon 6–6.

As an alternative to using a diamine and a dicarboxylic acid, an amino acid, such as $H_2N(CH_2)_5COOH$, can provide a polyamide. This amino acid forms a cyclic amide between its two ends. When this is heated, Nylon 6 is produced.

$$\xrightarrow{\text{heat}} \sim\!\sim NHCO(CH_2)_5NHCO(CH_2)_5NHCO \sim\!\sim$$

Nylon 6

An extremely useful polyamide, whose trade names include 'Kevlar' and 'Twaron', is made from a diamine and a dicarboxylic acid, both of which are derivatives of benzene (see also Chapter 27). This material has exceptional strength. Weight for weight it is five times stronger than steel and ten times stronger than aluminium. It is also heat-stable. In consequence it is a useful engineering material of high importance and can be used to replace metals in many applications, and for other purposes, e.g. in aircraft, tyres, cables, protective clothing, bullet-proof materials, asbestos replacement in brakes, and many other items where strength and heat resistance are important.

Proteins

Perhaps the most important of all polyamides are the proteins, used in multifarous ways by nature, as structural materials, as reagents, as catalysts, in processes such as biological reproduction and for countless other purposes. Examples of materials made from proteins are silk, wool, nails, hair, horn, hooves, skin, feathers, muscles, etc.

Proteins are condensation polymers made up from α-amino acids, $NH_2CHRCOOH$. Nature uses a wide variety of such amino acids, wherein R is not necessarily a simple alkyl group but may be a much more complicated structure. These amino acids join together to give polyamides, which are the protein:

etc. + NH$_2$CHRCOOH + NH$_2$CHR'COOH + NH$_2$CHR"COOH + etc.

$$\downarrow$$

etc. ∿ CO—NHCHRCO—NHCHR'CO—NHCHR"CO—NH ∿ etc.

The order in which the different amino acids link together is literally vital, and strictly controlled. With a large variety of possible constituent amino acids and an even larger number of ways of linking them together, nature shows enormous versatility in producing polyamides tailored to meet the numberless requirements. Proteins are an ultimate example of the vital importance of chemistry, and of the importance of chemical structure and chemical reactions.

Questions

1. How may propionyl chloride be converted into 1-aminopropane?
2. Give general formulae to illustrate the chemical structures of (a) polyesters, (b) polycarbonates and (c) polyamides. Indicate how they may be formed.
3. Draw equilibria to demonstrate that amides are amphoteric.

20 NATURALLY OCCURRING COMPOUNDS. CHIRALITY

Chemistry of Naturally Occurring Molecules

In some recent chapters a number of classes of compounds have been mentioned which are of great importance biologically in nature. Examples are proteins (Chapter 19), fats (Chapter 17), terpenes (Chapter 14) and carbohydrates (Chapter 9). These are only a few examples of the multitude of organic compounds, of very many different types and families, which are literally vital in that they play essential roles in the mechanism of life in humans, plants and animals.

Another important group of compounds is based on the **cyclopentanoperhydrophenanthrene** structure (this name will be explained in a later chapter):

Cyclopentanoperhydrophenanthrene

This structure consists of three six-membered rings and a five-membered ring all fused together. This formula does not give a good picture of the shape of this molecule since the rings are not flat (see Chapter 2); a three-dimensional picture looks as follows:

The shape of molecules is important and can play a crucial role in their chemistry.

Many **steroids** are based on this structure. Biologically important compounds include the D-vitamins (which prevent rickets and promote bone and teeth formation), cholesterol, cortisone, cardiac stimulants and sex hormones. For example, testosterone and androsterone, shown below, are among the compounds responsible for the development of male characteristics in human beings; estrone and progesterone are female sex hormones. The relatively small differences between the male and female hormones is noteworthy.

Testosterone

Androsterone

Estrone

Progesterone

As has been mentioned earlier, in general terms the chemistry of these complicated compounds can be forecast from a knowledge of the chemistry of the functional groups present in the molecules. Thus androsterone will give reactions typical of a hydroxy group and of a ketone. Problems can arise when two (or more) functional groups are adjacent to one another. This is the case for testosterone, where, because the carbonyl and alkene groups are next to one another, their properties are somewhat modified from those of isolated carbonyl and alkene groups. Interaction between groups in the same molecule has always to be looked for, and allowed for, but with due care for this, properties of more complex molecules can be forecast from knowledge of simpler molecules. The shape of molecules can also play an important role in their chemistry, by enabling groups to interact with reagents more readily or by hindering such interaction.

Chirality

One facet of chemical structure and reactivity, which is highly important for biological molecules but which has not yet been mentioned in this text, is that more complex molecules may have a non-symmetrical structure, and in consequence can exist in structures of opposing symmetry, i.e. in right-handed and left-handed forms. Such molecules are described as **chiral**, a word derived from the Greek word for a hand, *cheir*. **Chirality** is the phenomenon of handedness.

This concept is not only chemical, it is very important in everyday life. We have right and left hands and feet, right and left gloves and shoes; screws and propellors are right- and left-handed.

It is characteristic of pairs of right- and left-handed objects that:

1. The objects are identical in all respects except their handedness.
2. The one form is the mirror-image of the other; if you look at a right-handed object in a mirror it appears as the left-handed equivalent.
3. It is impossible to pack a right-handed form exactly on top of its left-handed equivalent; the right- and left-handed forms are said to be not superimposable.

Very many common objects are chiral, e.g. grains of quartz, grand pianos and flatfish.

The only experimental way to distinguish between chiral forms is by their interaction with other chiral objects. An **achiral** (non-chiral) object, if it is the right size, will fit equally well into a right or left shoe; it cannot distinguish between them. However, a right shoe will only fit properly onto a right foot and a left shoe onto a left foot. The same applies to hands and gloves.

Molecules can be chiral if their molecular structure is not symmetrical. The most simple cases arise for molecules in which four different groups are attached to one carbon atom, as, for example, in 2-bromobutane:

$$CH_3 - \underset{\underset{H}{|}}{\overset{\overset{Br}{|}}{C}} - CH_2CH_3$$

Attached to the 2-carbon atom are a hydrogen atom, a bromine atom, a methyl group and an ethyl group. If this molecule is drawn three dimensionally or, better, if a molecular model of it is made, it is found that there are two different ways of doing this, as illustrated in formulae **A** and **B**:

                (A)   (B)

Compounds **A** and **B** are not identical. They fulfil the requirements of chiral pairs as listed above. The molecules are mirror images of each other, and they are not superimposable on one another. Because they are different molecules they cannot be changed into one another without breaking and remaking bonds in the process. The central carbon atom to which the different groups are attached is commonly described as a **chiral carbon atom** or an **asymmetric carbon atom**, (although strictly it is not the carbon atom itself, but its environment, which is asymmetric or chiral; it is an **asymmetrically substituted** carbon atom).

As happens in the interactions between feet and shoes, it may be that one chiral isomer can react satisfactorily with only one of two chiral forms of another molecule. As a purely hypothetical example let us consider a reaction between two different compounds QCXYZ and RCXYZ, and let us further suppose that for the reaction to proceed it is essential that the groups X, Y, Z of QCXYZ can, at the same time, meet, respectively, groups X, Y and Z of RCXYZ. From the diagram below it is evident that for one chiral isomer of QCXYZ, reaction will be possible with only one of the chiral isomers of RCXYZ, but not with the other:

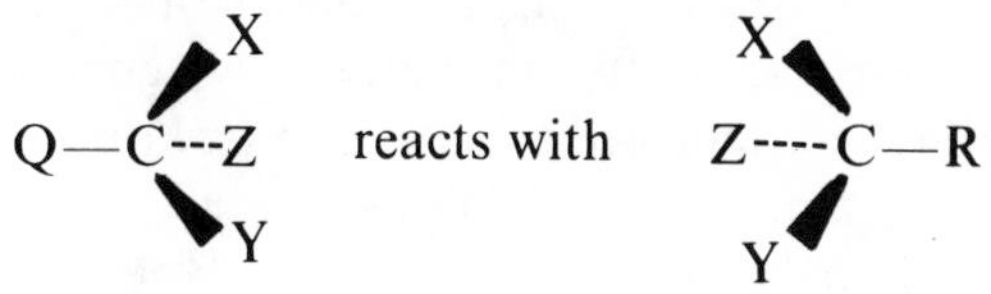

$$Q—C\underset{Y}{\overset{X}{\cdots Z}} \quad \text{reacts with} \quad Z\cdots\underset{Y}{\overset{X}{C}}—R$$

but

$$Q—C\underset{Y}{\overset{X}{—Z}} \quad \text{cannot react with} \quad Z\cdots\underset{X}{\overset{Y}{C}}—R$$

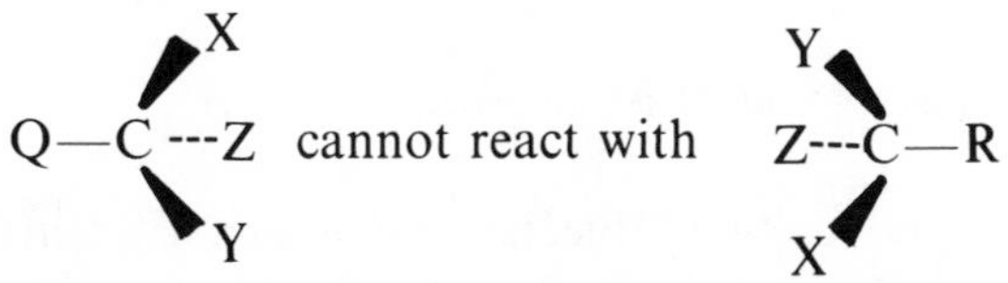

Nature makes great use of this specificity. Since so many molecules involved in biological reactions are chiral, their reactions are often limited to those with other molecules of the appropriate chirality. (A very human parallel may be drawn to a thirsty right-handed man trying to draw the cork out of a bottle with a left-handed corkscrew.)

Consequences of this chiral selectivity may be seen in the very different physiological properties shown by pairs of chiral isomers. A few common examples selected from the huge range available are as follows:

1. One chiral form of a molecule carvone is responsible for the characteristic taste of caraway; the other chiral form tastes of spearmint.
2. One chiral form of nicotine, that found in tobacco, is highly toxic; the other chiral form is not.
3. One chiral form of ascorbic acid is the essential vitamin, vitamin C; the other chiral form is ineffective.
4. One chiral form of thalidomide is responsible for the tragedies brought about by its use; the other chiral form is apparently a beneficent drug.

Most of the molecules taking part in the chemistry of the human body are chiral. For example, the proteins are polyamides made up from chiral amino acids:

$$\begin{array}{c} R \\ NH_2CCOOH \\ H \end{array}$$

of particular chirality. Note that the central carbon atom in this formula has four different groups attached to it, H, R, NH_2, COOH. If a space traveller landed in another world, apparently identical to Earth, but in fact made up, and working on, molecules of opposite chirality, its food would be largely useless to him because his enzymes would be unable to digest it. If he attempted to have a child with a woman of the other planet, it would prove to be impossible because of the chiral incompatibility of the two beings.

Special Properties of Chiral Molecules

As mentioned above, two objects of opposite chirality which are otherwise identical interact in exactly the same way with achiral reactants or with an achiral environment. In consequence, properties of two chiral forms of a chemical compound, such as melting points, solubility in achiral solvents and reactions with achiral molecules, are identical for the two forms.

However, this changes when interactions with other chiral molecules are involved. Two chiral forms of a molecule have different solubilities in a chiral solvent, and, with chiral reagents, will react at different rates or in fact only one of the chiral forms may take part in a reaction.

Chirality is usually identified by interaction of a molecule with plane-polarized light. Plane-polarized light is light in which all the wave vibrations have been filtered out except for those in one particular plane. It is generated from ordinary light by passing it through either a suitably constructed prism made from calcite (calcium carbonate) or a Polaroid lens. If this plane-polarized light is passed through a second calcite prism or Polaroid lens, they can also filter out all light except that with wave vibrations in one particular plane. If the first and second prisms (or lenses) are suitably aligned, the plane-polarized light that passes through the first prism also passes through the second prism. If the prisms (or lenses) are not so aligned, the second prism filters out the light from the first prism and no light passes through the second prism. This can be observed by looking through two Polaroid lenses held one in front of the other. When they are aligned in a certain way light passes through both, but when one of them is rotated in its own plane the light is cut off.

This can be observed in a **polarimeter**, diagrammatically shown below:

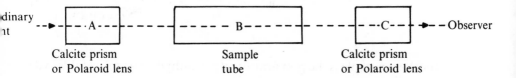

Calcite prism   Sample   Calcite prism
or Polaroid lens   tube   or Polaroid lens

When prisms A and C are appropriately aligned light passes through both and is detected by the observer. If sample tube B is empty, or contains an achiral sample, no adjustment needs to be made to the alignment of A and C for light to be seen by the observer. If, however, a chiral sample is put into B, then it is necessary to rotate C from its original setting in order to observe light. This means that the sample B must have rotated the plane of vibration of the light waves passing through it and as a result C, as originally set up, filters this light off. If, however, C is rotated until it is suitably aligned with the plane-polarized light which is reaching it, then that light passes through C. (If the simple experiment mentioned above, involving looking through two Polaroid lenses, is repeated with a solution of sugar held between the two lenses, it will be found necessary to rotate the second lens to a different position, because sugar is chiral.)

It is found that equal quantities of the two chiral forms of one compound rotate the plane of polarization of light to the same extent, but one rotates it in one direction and the other in the opposite direction. The symbolism of d or $+$ is used if the prism C has to be rotated to the right and l or $-$ if the prism has to be rotated to the left, to enable the light to be transmitted through it. A 1:1 mixture of d and l forms of a particular compound cancel out the rotations, and no rotation of C is needed. Different molecules rotate the plane of polarization of light by different extents.

When chiral molecules are prepared in the laboratory from achiral precursors, a 1:1 mixture of the two forms is normally produced. This is because a reagent normally has an equal chance of reacting from either side. This may be illustrated by considering the formation of 2-bromobutane from but-2-ene:

Another example is the preparation of a cyanhydrin from acetaldehyde:

Mirror-image forms

If the two chiral forms of acetaldehyde cyanhydrin are converted **separately** into lactic acid (CH_3—CHOH—COOH) then chiral lactic acids will be formed:

Mirror-image forms of lactic acid

Conversion of the cyano group into a carboxyl group does not involve the central chiral carbon atom in the reaction, and hence it retains its chirality.

In nature, reactions often start with chiral reactants and give chiral products. Nature also introduces chirality in reactions by using chiral catalysts to influence reactions of achiral starting materials. These catalysts promote the preferred formation of one chiral form.

In Chapter 7, the S_N1 and S_N2 mechanisms of nucleophilic substitution reactions were considered. If a chiral reactant, such as 2-

bromobutane, is used in an S_N2 type reaction the chirality is reversed from right handed to left handed (or vice versa) in the reaction, because of the approach of the nucleophile from the opposite side of the molecule to that from which the bromine atom leaves:

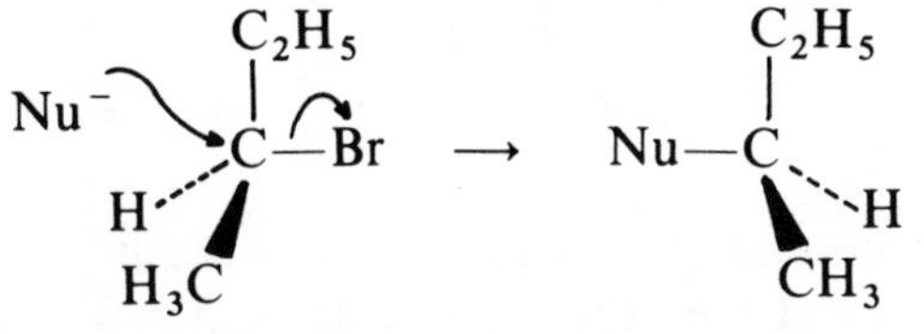

$$(Nu^- = nucleophile)$$

In an S_N1 type reaction one chiral from gives rise to both chiral forms of the product, because a planar intermediate is formed, which can then be attacked by the nucleophile from either side:

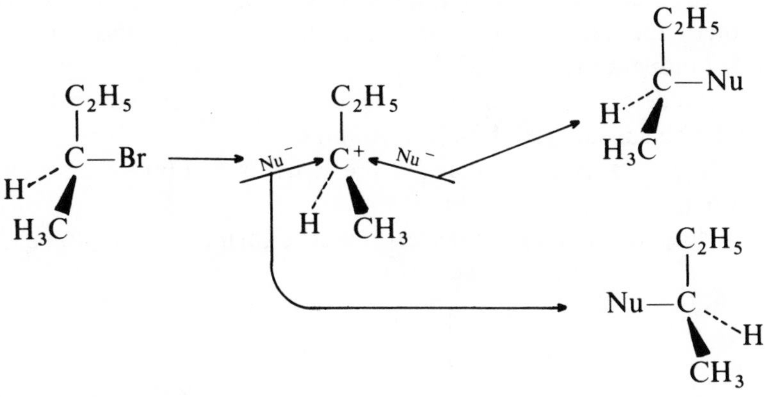

It is thus possible to investigate the mechanism of a nucleophilic substitution reaction involving a chiral compound by observing whether chirality is reversed in the product or whether a mixture of both chiral forms results.

The presence of a carbon atom with four different groups attached to it is not the only source of chirality in a molecule, although it is perhaps the commonest cause. A whole molecule may be chiral rather than any specific atom. An example of such chirality is provided by the following diene, whose shape is as shown below and for which two mirror-image forms are possible:

The groups at one end of the two double bonds lie in the plane of the paper; the groups at the other end lie at right angles to this plane.

Questions

1. Which of the following compounds have chiral carbon atoms? Identify the chiral atoms in the compounds which have them. (a) 2-chlorobutane, (b) 2-methylbutane, (c) 3-methylpentane, (d) 3-chloro-2-methylpentane, (e) 1-bromo-1-chlorobutane and (f) 2,3-dibromopentane.
2. Write a detailed mechanism of the reaction between sodium cyanide and a chiral sample of 2-chlorobutane.
3. Write generalized formulae for the structures of (a) a fat and (b) a protein.
4. List five reactions of cholesterol, whose formula is as shown:

   it is not necessary to write out the full structure every time; just indicate the relevant part, for example

5. Identify all the functional groups in the following molecules. C_6H_5 is the **phenyl** group, which will be encountered in later chapters (Chapters 24 to 27).

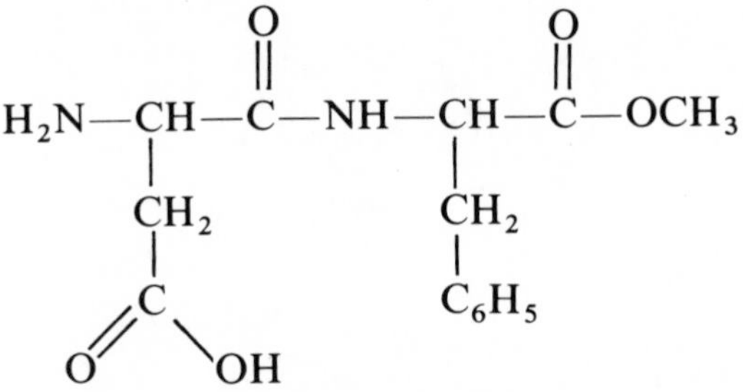

$$H_2N-\underset{\underset{\underset{\underset{O}{\diagup\diagdown}OH}{C}}{CH_2}}{CH}-\overset{\overset{O}{\|}}{C}-NH-\underset{\underset{C_6H_5}{CH_2}}{CH}-\overset{\overset{O}{\|}}{C}-OCH_3$$

Aspartame
(a sweetening agent
for foodstuffs)

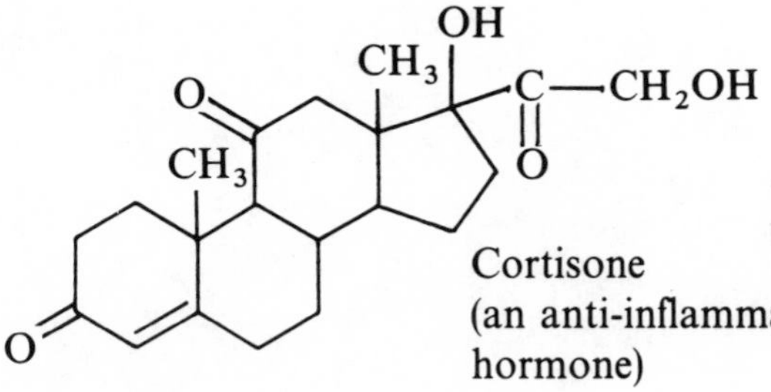

Cortisone
(an anti-inflammatory
hormone)

21 COMPOUNDS WITH CARBON–NITROGEN MULTIPLE BONDS

Imines

Compounds containing carbon–carbon double bonds (alkenes) and compounds containing carbon–oxygen double bonds (carbonyl compounds) have been considered. The analogous compounds containing carbon–nitrogen double bonds are known as **imines**:

$$\begin{array}{c} R \\ \diagdown \\ C\!=\!N \\ \diagup \\ R \end{array}\!\!{}^{\diagup H} \qquad \text{or} \qquad \begin{array}{c} R \\ \diagdown \\ C\!=\!N \\ \diagup \\ R \end{array}\!\!{}^{\diagup R'}$$

They are formed by the reaction, usually acid catalysed, of aldehydes or ketones with ammonia or substituted derivatives of ammonia. This reaction has been discussed in Chapter 14.

Imines in which R and R' in the formulae above are alkyl groups are readily hydrolysed. The carbon–nitrogen double bond, like the carbon–oxygen double bond, is polarized, and with water undergoes a reaction that is the reverse of the formation of imines from carbonyl compounds:

$$\begin{array}{c} R \\ \diagdown \\ C\!=\!NR' \\ \diagup \\ R \end{array} \xrightarrow{\text{H}_2\text{O}} \begin{array}{c} R \\ \diagdown \\ C\!=\!O \\ \diagup \\ R \end{array} + \text{H}_2\text{NR}'$$

Imines which have an NH group also readily polymerize on standing.

If the two groups attached to the carbon atom are different, geometric isomers are possible:

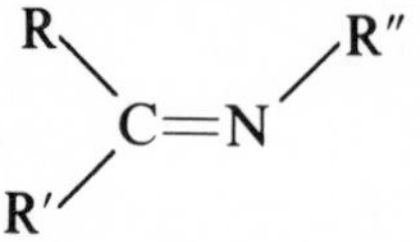

$$\underset{R'}{\overset{R}{\diagdown}}C{=}N\diagup^{R''} \quad \text{and} \quad \underset{R'}{\overset{R}{\diagdown}}C{=}N\underset{\diagdown R''}{}$$

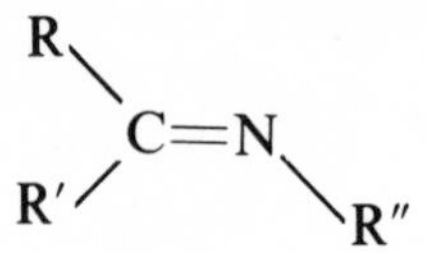

(cf. alkenes; see Chapter 3).

Cyanides or Nitriles

The preparation of **cyanides** or **nitriles** from alkyl halides was mentioned earlier:

$$R{-}Cl \xrightarrow{\ ^-CN\ } R{-}CN$$

These compounds contain a **triple** bond between the carbon and nitrogen atoms, $R{-}C{\equiv}N$.

They are named either as cyanoalkanes or by reference to the related carboxylic acid, $R{-}COOH$, for example:

CH_3CN Methyl cyanide or cyanomethane or acetonitrile (from acetic acid)

$CH_3CH_2CH_2CN$ propyl cyanide or 1-cyanopropane or butanenitrile (from butanoic acid)

The $C{\equiv}N$ bond is polarized in the same sense as a carbonyl or imine group:

$$R{-}\overset{\delta+}{C}{\equiv}\overset{\delta-}{N}$$

Two important reactions of nitriles are their hydrolysis and their reduction.

Hydrolysis may be catalysed by acid or by base, as follows:

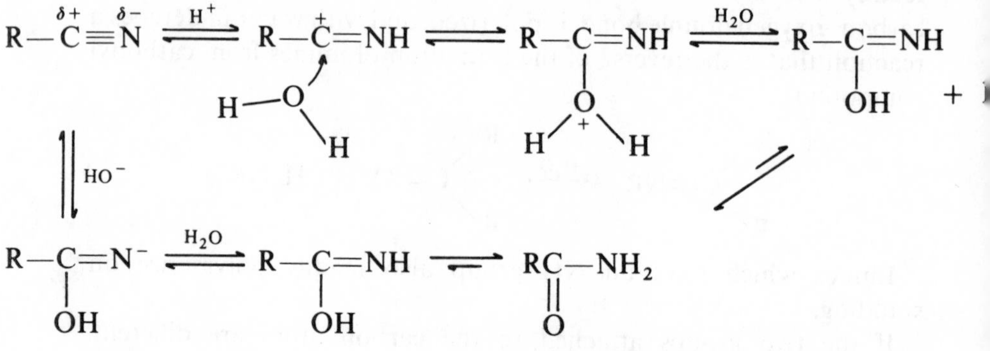

An amide

The amide that is formed may then be hydrolysed further to a carboxylic acid, so that the overall reaction is conversion of the nitrile into a carboxylic acid:

$$RCN \xrightarrow[\text{H}^+ \text{ or HO}^-]{\text{H}_2\text{O}} RCOOH$$

Since nitriles are readily prepared from alkyl halides, this is a useful synthetic route to carboxylic acids.

Nitriles can be reduced to amines, using either lithium aluminium hydride, or hydrogen in the presence of a catalyst:

$$RCONH_2 \xrightarrow[\text{or H}_2/\text{catalyst}]{\text{LiAlH}_4} RCH_2NH_2$$

Their real importance chemically is in providing a means of making a new carbon–carbon bond, and thus adding an extra carbon atom to a carbon chain:

$$RCl \longrightarrow RCN \begin{array}{l} \nearrow RCOOH \\ \searrow RCH_2NH_2 \end{array}$$

Alkyl cyanides are less toxic than cyanide salts because they are covalently bound and do not provide cyanide ions, which are extremely toxic.

Questions

1. How may acetonitrile be converted into acetamide? Give mechanisms for the reactions.
2. Suggest a mechanism for the conversion of an imine $R_2C{=}NR^1$ into a ketone R_2CO by hydrolysis with aqueous acid.
3. How could ethylamine be prepared, starting from bromomethane?

22 ALKYNES

In the foregoing chapter compounds having carbon–nitrogen double and triple bonds were considered. Alkenes have been discussed earlier (Chapters 3 and 5). Now we turn to compounds with carbon–carbon triple bonds. These compounds are called **alkynes**. The simplest member of the series, $HC\equiv CH$, is almost always known by its trivial but official name **acetylene**. Higher members of the series are named from the related alkane by replacing the ending '-ane' by '-yne', for example

$CH_3-C\equiv CH$	Propyne
$CH_3CH_2C\equiv CH$	But-1-yne
$CH_3C\equiv CCH_3$	But-2-yne

The general formula of the acyclic alkynes is C_nH_{2n-2}.

It was seen in Chapter 3 that carbon atoms joined together by a double bond are closer together than those joined by a single bond. The triple bond is even shorter, as illustrated by the following carbon–carbon bond lengths:

In ethane	H_3C-CH_3	1.53 Å
In ethylene	$H_2C=CH_2$	1.34 Å
In acetylene	$HC\equiv CH$	1.21 Å

In theory alkynes could be prepared from 1,1- or 1,2-dibromoalkanes, by the action of base:

In fact these reactions do not work very satisfactorily and in any case require a very strong base, e.g. sodamide, $Na^+ NH_2^-$. The difficulty lies in removing the second HBr, which is much more difficult to achieve from a bromoalkene than it is from a bromoalkane.

Higher alkynes are commonly made from acetylene, as discussed below.

Acetylene

Acetylene is usually made industrially by cracking oil fractions at high temperatures. Especially it can be made by heating methane very strongly for a very short period of time:

$$CH_4 \xrightarrow[<0.1\,s]{1500\,°C} HC\equiv CH$$

If the heating is continued longer the acetylene decomposes. Another source of acetylene is from the cracking of ethylene.

The older method, now only used residually to a small extent, consisted of heating coke and lime together in an electric furnace at a very high temperature to form **calcium carbide**, CaC_2, which reacts with water to form acetylene:

$$CaO + 3\,C \xrightarrow{heat} Ca^{2+} C_2^{2-} \xrightarrow{H_2O} HC\equiv CH$$
$$+ CO$$

Acetylene is an important industrial chemical. It is a gas at room temperature. It is not very stable as a gas, nor when liquified. Under pressure, or in the presence of some catalysts, e.g. copper, it can decompose with explosive violence to give carbon and hydrogen. It is stored and transported in cylinders as a solution in acetone absorbed on pumice.

It has a very large use in cutting and welding of metals, since it burns with a very hot flame ($\sim 2800\,°C$).

At one stage it was a very important industrial chemical intermediate, but it is much less used now. Its main use is for making a polymer named **neoprene**, which is used as a synthetic rubber; it has considerable resistance to oils, to oxygen and to heat:

$$HC\equiv CH \xrightarrow{\text{catalyst}} HC\equiv C-CH=CH_2$$

$$\xrightarrow{\text{HCl}} CH_2=CCl-CH=CH_2$$

$$\xrightarrow{\text{polymerize}}$$

Neoprene

Addition Reactions

Like alkenes, alkynes are unsaturated and electron-rich, and undergo addition reactions with electrophiles similar to those of alkenes. Examples are

$$HC\equiv CH \begin{cases} \xrightarrow{Br_2} HBrC=CHBr \xrightarrow{Br_2} HBr_2C-CHBr_2 \\ \xrightarrow[Ni]{H_2,} CH_3-CH_3 \\ \xrightarrow{H_2 \text{ with specially deactivated catalyst}} CH_2=CH_2 \end{cases}$$

Acids, HX, also add to acetylene, in general

$$HC\equiv CH + HX \longrightarrow H_2C=CHX$$

Since this product is an alkene it can add a further molecule of HX.

The grouping $H_2C=CH-$ is known as the **vinyl** group; for example $H_2C=CHCl$ is **vinyl chloride** (or, alternatively, chloroethylene).

Some of these addition reactions used to be carried out industrially, but the products are now for the most part made by other routes. Such processes were

$$HC{\equiv}CH \xrightarrow{}$$

$$\overset{HCl}{\nearrow} H_2C{=}CHCl$$
Vinyl chloride

$$\xrightarrow{HCN} H_2C{=}CHCN$$
Acrylonitrile ($H_2C{=}CHCOOH$ is acrylic acid)

$$\overset{CH_3COOH}{\searrow} CH_3COOCH{=}CH_2$$
Vinyl acetate

Vinyl chloride is polymerized to give **poly(vinyl chloride)**, used for insulation, piping, floor tiles, records (discs) and many other uses. Acrylonitrile is converted into **poly(acrylonitrile)**, which is used as a synthetic fibre. Vinyl acetate is converted into **poly(vinyl acetate)**, used in adhesives, paints, etc. A mixture of poly(vinyl chloride) (PVC) and poly(vinyl acetate)(PVA) is used for floor coverings, furniture coverings and soles of shoes.

It is easier to produce alkenes than alkynes, and so alkenes are now the usual source of these products. For example, vinyl chloride and acrylonitrile are made from, respectively, ethylene and propene:

$$CH_2{=}CH_2 \xrightarrow{Cl_2} CH_2Cl{-}CH_2Cl \xrightarrow{-HCl} CH_2{=}CHCl$$

$$CH_3CH{=}CH_2 + O_2 + NH_3 \xrightarrow{catalyst} CH_2{=}CHCN$$

Another formerly important addition process was the conversion of acetylene into acetaldehyde:

$$HC{\equiv}CH \xrightarrow[\substack{HgSO_4 \\ H_2SO_4}]{H_2O} CH_3{-}CHO$$

As mentioned above, acetylene is still used in the manufacture of neoprene. The first step here is a catalysed addition of one molecule of acetylene to a second molecule of acetylene.

The other processes might well be revived at a future date if economics dictate that acetylene becomes a cheaper source than alkenes for making the products involved.

Acidity of Acetylene

A carbon atom linked by a triple bond tends to hold all the electrons around it more tightly than a saturated carbon atom does. A consequence of this is that a hydrogen atom in a $\equiv$CH group is very weakly acidic; since the electrons of the C—H bond are more associated with the carbon atom, release of the hydrogen atom as a proton is easier to achieve. The $\equiv$C—H group is still only very weakly acidic. It is a weaker acid than water, but strong bases such as sodamide can form a sodium salt. This applies not only to acetylene itself but to all alkynes of general formula RC$\equiv$CH having a hydrogen atom attached to one end of the triple bond:

$$RC\equiv CH \xrightarrow{\text{NaNH}_2} RC\equiv C^- Na^+$$

for example

$$HC\equiv CH \xrightarrow{\text{NaNH}_2} HC\equiv C^- Na^+$$

Sodium acetylide

Since water is a stronger acid than are alkynes, salts of the latter react with water, which reconverts them into the alkyne

$$RC\equiv C^- Na^+ \xrightarrow{\text{H}_2\text{O}} RC\equiv CH + Na^+ \, {}^-OH$$

In days before the use of spectroscopic techniques for identification purposes, alkynes containing the $\equiv$C—H group, i.e. so-called **terminal alkynes**, with the alkyne group at the end of the carbon chain, were detected by means of the copper and silver salts which they formed:

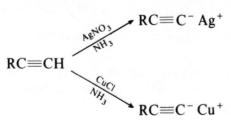

These salts separate out as grey and reddish coloured precipitates respectively. Their formation provides evidence for the presence of the $\equiv$CH grouping. These salts, like other heavy-metal salts of alkynes, are explosive when dry.

As salts of weak acids, the alkyne anions are strong nucleophiles. Hence they react readily with alkyl halides:

$$Na^+\ RC\equiv C^- \quad R'-Br \longrightarrow RC\equiv CR' + Br^-\ Na^+$$

This provides a way of making higher members of the alkyne series.

Acetylene itself can be converted into monoalkyl or dialkyl alkynes, for example

$$Na^+\ HC\equiv C^- \quad CH_3-Br \longrightarrow HC\equiv CCH_3 \quad \text{Propyne}$$

$$\Big\downarrow NaNH_2$$

$$CH_3C\equiv CCH_3 \longleftarrow Br-CH_3 \longleftarrow ^-C\equiv CCH_3$$
But-2-yne

As nucleophiles the alkyne salts can also react with carbonyl compounds to give alcohols:

$$Na^+RC\equiv C^- \quad \begin{array}{c} R' \\ \diagdown \\ C \\ \| \\ O \end{array} \diagup R' \longrightarrow RC\equiv C-\overset{\displaystyle R'}{\underset{\displaystyle O^-}{\overset{\displaystyle |}{\underset{\displaystyle |}{C}}}}-R'$$

$$\diagdown H_2O$$

$$RC\equiv C-CR_2'OH$$

Because of the acidity of terminal alkynes they react with Grignard reagents to form alkynyl Grignard reagents, for example

$$CH_3C\equiv CH + C_2H_5MgBr \longrightarrow CH_3C\equiv CMgBr + C_2H_6$$

This is another example of the standard reaction of Grignard reagents with acids to form alkanes.

These alkynyl Grignard reagents undergo the usual reactions of Grignard reagents, e.g. they react with ketones to form tertiary alcohols, for example

$$CH_3C{\equiv}CMgBr + CH_3COCH_3 \rightarrow$$

$$CH_3C{\equiv}C{-}\underset{\underset{OMgBr}{|}}{C}(CH_3)_2 \xrightarrow[H_2O]{H^+} CH_3C{\equiv}C{-}\underset{\underset{OH}{|}}{C}(CH_3)_2$$

This method of making alcohols from alkynes has some advantages over the alternative method using the sodium salt of the alkyne.

Questions

1. Three different isomers have molecular formula C_4H_6; each shows two signals in its ^{13}C-n.m.r. spectrum. Two of these isomers show two signals and the third shows one signal in their ^1H-n.m.r. spectra. Write structural formulae for these compounds.
2. What products do you obtain from the reaction of acetylene with (a) bromine, (b) hydrogen chloride, (c) aqueous sulphuric acid and mercury(II) sulphate and (d) hydrogen in the presence of nickel? Does the product from (d) have a longer or shorter bond than acetylene?
3. How could you distinguish between but-1-yne and but-2-yne, (a) spectroscopically and (b) by chemical tests?
4. How might the following compounds be prepared from acetylene: (a) acetic acid, (b) α-hydroxypropionic acid, (c) pent-1-yne? (More than one step is required in each case.)
5. How might 4-hydroxy-4-methylpent-2-yne be obtained, starting from propyne and acetone?

23 DIENES AND POLYENES

Alkynes, which were dealt with in the previous chapter, have the general formula C_nH_{2n-2}. This same general formula may also represent another class of hydrocarbons, the **dienes**, i.e. molecules with two carbon–carbon double bonds (and also cyclic molecules with one double bond such as cyclopentene, C_5H_8). All of these classes of compounds are thus isomeric with each other.

Dienes

Dienes are often classified into three types, depending upon the relative placing of the two double bonds. These classes may be illustrated by the isomeric 1,2-, 1,3- and 1,4- pentadienes.

CH_2=CH—CH_2—CH=CH_2 **Isolated** double bonds
Penta-1,4-diene

CH_2=CH—CH=CH—CH_3 **Conjugated** double bonds
Penta-1,3-diene

CH_2=C=CH—CH_2—CH_3 **Cumulated** double bonds
Penta-1,2-diene   or **Allenes**

In dienes with **isolated** double bonds, the double bonds are separated from one another by more than one single bond. They behave like simple alkenes and there is little interaction, if any, between the two double bonds.

Allenes, which have adjacent, or cumulated, double bonds, are not very common compounds and have rather special properties. They will not be considered in this text. Their shape is briefly mentioned at the end of Chapter 20.

Conjugated double bonds are separated from one another by one single bond, and are so called because the two double bonds interact with one another and affect each other's properties. This is a further example of the effects of neighbouring groups upon one another, as noted a number of times previously in this text.

The maximum absorption in the ultraviolet spectra of conjugated dienes is more intense and occurs at longer wavelengths than in the case of isolated double bonds. This can be used to show that double bonds in a diene are conjugated.

Addition Reactions

When 1 mol of hydrogen chloride adds to 1 mol of buta-1, 3-diene, $CH_2=CH-CH=CH_2$, a mixture of two different products results:

$$CH_2=CH-CHCl-CH_3$$

$$CH_2=CH-CH=CH_2 \xrightarrow{1HCl} \text{3-Chlorobut-1-ene}$$

$$+$$

$$CH_2Cl-CH=CH-CH_3$$
1-Chlorobut-2-ene

How does this come about?

In each product a hydrogen atom has added to a terminal carbon atom. Let us presume that, as in the case of simple alkenes, this is the first step of the reaction:

$$CH_2=CH-CH=CH_2 \quad H-Cl \longrightarrow CH_2=CH-\overset{+}{C}H-CH_3 + Cl^-$$

In the carbenium ion which is formed there is a situation comparable to that in a carboxylate anion. In the latter, negative charge can be spread, or **delocalized**, by electronic interaction with the adjacent carbonyl group:

$$-\overset{O}{\underset{O^-}{C}} \longleftrightarrow -\overset{O^-}{\underset{O}{C}} \quad \text{or} \quad -\overset{\overset{\delta-}{O}}{\underset{\underset{\delta-}{O}}{C}}$$

In the cation formed by addition of a proton to a conjugated diene, the positive charge can similarly be spread or delocalized by interaction with the 'extra' electrons present in the other double bond:

$$CH_2 \!=\! CH \!-\! \overset{+}{C}H \!-\! CH_3$$

$$\text{or} \quad \overset{\delta+}{C}H_2 \!-\! CH \!-\! \overset{\delta+}{C}H_2 \!-\! CH_3$$

$$\overset{+}{C}H_2 \!-\! CH \!=\! CH \!-\! CH_3$$

In this case two electrons are shared over three carbon atoms. The central carbon atom is associated with eight electrons; the carbon atoms at either side have a lesser share in these electrons and hence carry positive charge.

This is why the proton adds to a terminal carbon atom of a conjugated diene. Were it instead to add to a non-terminal carbon atom, the resultant positively charged carbon atom would be isolated by a CH_2 group from the remaining double bond:

$$CH_2 \!=\! CH \!-\! CH_2 \!-\! \overset{+}{C}H_2$$

Hence the electrons of the double bond could not interact with the positively charged carbon atom and spread out the charge; it has been noted earlier in this text that spreading out of charge lowers the energy of an ion (or molecule) and hence makes it more stable. An energy profile can be constructed to illustrate the preferential addition of a proton to the end atom of a diene, as follows:

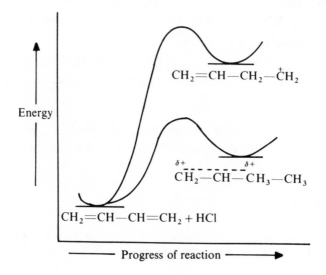

The second step of the overall reaction involves addition of chloride ion to the intermediate carbenium ion. This might happen at either of the positively charged carbon atoms, to give the observed products:

$$\overset{\delta+}{\text{C}}\text{H}_2 \text{---CH}\text{---}\overset{\delta+}{\text{CH}}\text{---CH}_3 \xrightarrow{\text{Cl}^-} \text{CH}_2\text{==CH---CHCl---CH}_3$$

$$\text{1, 2-Addition}$$

$$\text{Cl}^- + \overset{\delta+}{\text{CH}}_2\text{---CH}\text{---}\overset{\delta+}{\text{CH}}\text{---CH}_3 \rightarrow \text{CH}_2\text{Cl---CH==CH---CH}_3$$

$$\text{1, 4-Addition}$$

The two modes of addition are described, respectively, as 1, 2- and 1, 4-addition.

A similar situation arises with the addition of halogens to conjugated dienes, for example

$$\text{CH}_2\text{==CH---CH==CH}_2 + \text{Br---Br} \rightarrow \overset{\delta+}{\text{C}}\text{H}_2\text{---CH}\text{---}\overset{\delta+}{\text{CH}}\text{---CH}_2\text{Br} + \text{Br}^-$$

$$\text{Br}^- \qquad\qquad\qquad \text{Br}^-$$

CH$_2$Br—CH=CH—CH$_2$Br   CH$_2$=CH—CHBr—CH$_2$Br

1, 4-Bromobut-2-ene   3, 4-Dibromobut-1-ene

If more bromine is present then two more atoms of bromine can add to either of these dibromobutenes to give a tetrabromobutane:

CH$_2$Br—CH=CH—CH$_2$Br $\xrightarrow{Br_2}$

$\xrightarrow{Br_2}$ CH$_2$Br—CHBr—CHBr—CH$_2$Br

CH$_2$=CH—CHBr—CH$_2$Br   1, 2, 3, 4-Tetrabromobutane

Dienes, like simple alkenes, can add on hydrogen in the presence of a suitable catalyst to give, in turn, an alkene and an alkane.

Cyclo-addition Reactions

Another special reaction of conjugated dienes is an addition reaction that leads to the formation of a six-membered ring. An example is

$$
\begin{array}{c}
\overset{\displaystyle CH_2}{\overset{\parallel}{CH}} \\
\underset{\displaystyle CH_2}{\overset{\mid}{CH}}
\end{array}
\quad + \quad
\begin{array}{c}
\overset{\displaystyle CHO}{CH} \\
\underset{\displaystyle CH_2}{\parallel}
\end{array}
\quad \longrightarrow \quad
\begin{array}{c}
CH_2 \\
CH \quad\quad CHCHO \\
\parallel \quad\quad\quad \mid \\
CH \quad\quad CH_2 \\
CH_2
\end{array}
$$

In this reaction the other reactant is an alkene with a substituent group which tends to draw electrons away from the carbon–carbon double bond by conjugation with it, i.e. it is a so-called **electron-withdrawing** substituent. In the present example this can be pictured to happen as follows:

$$
\left[
\begin{array}{c}
H \\
\mid \\
C = O \\
CH \\
\parallel \\
CH_2
\end{array}
\longleftrightarrow
\begin{array}{c}
H \\
\mid \\
C - O^- \\
CH \\
\mid \\
+ CH_2
\end{array}
\right]
\quad \text{or} \quad
\begin{array}{c}
H \\
\mid \\
C - \overset{\delta-}{O} \\
HC \\
\parallel \\
CH_2 \\
{\scriptstyle \delta+}
\end{array}
$$

The alkene bond of this reactant adds to the terminal atoms of the diene in a 1,4-addition reaction.

Reactions of this sort, in which two molecules combine together to form a cyclic compound, are called **cyclo-addition reactions**.

The particular example given above, namely the addition of a 1,3-diene to an alkene having an electron-withdrawing substituent to form a cyclohexene derivative, is known as the **Diels–Alder reaction**, after the two chemists who developed and exploited its possibilities. It is an extremely important and useful synthetic reaction.

Its mechanism can be pictorially represented as follows:

$$
\text{(diene)} \quad + \quad \text{CHO}
$$

The six electrons of the alkene bonds, four from the diene and two from the other reactant, all interact with one another to form the six-membered ring. It will be seen later that the number of electrons, six, is important.

Some molecules other than alkenes which can provide two electrons may react similarly with dienes, for example

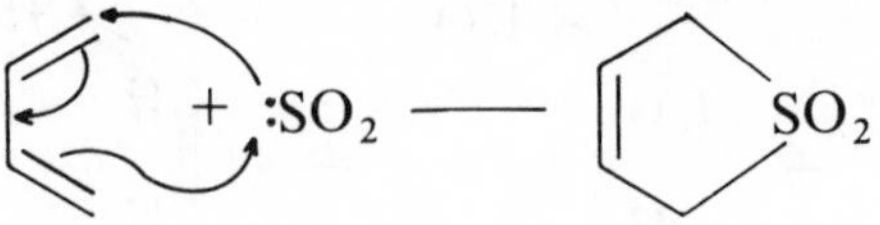

This reaction is used as an industrial process.

Polymerization

Dienes polymerize readily. The resultant polymer has double bonds at intervals along its molecular chain. These polymers are of great industrial importance.

Poly(butadiene) is made industrially on a very large scale by radical-catalysed polymerization of butadiene. Reaction proceeds as follows (cf. Chapter 5):

$$X^\cdot \quad CH_2 = CH - CH = CH_2 \rightarrow XCH_2 - CH = CH - \dot{C}H_2$$

$X^\cdot$ = Initiator

$$XCH_2 - CH = CH - CH_2 - CH_2 - CH = CH - \dot{C}H_2 \leftarrow CH_2 = CH - CH = C$$

etc. etc.

$$+CH_2 - CH = CH - CH_2 +_n$$

Poly(butadiene)

Addition of the dienes is 1, 4. The poly(butadiene) is used as a synthetic rubber, and especially is used in tyres.

Natural rubber is a polymer of 2-methylbuta-1, 3-diene, generally known as **isoprene**:

$$CH_2 = \underset{\underset{CH_3}{|}}{C} - CH = CH_2 \qquad \qquad +CH_2 - \underset{\underset{CH_3}{|}}{C} = CH - CH_2 +_n$$

Isoprene   Natural rubber

Another industrial polymer made from a diene is neoprene, which is made by polymerization of **chloroprene** (see Chapter 22):

$$CH_2\!=\!\overset{\overset{\displaystyle Cl}{|}}{C}\!-\!CH\!=\!CH_2 \qquad\qquad \left(\!CH_2\!-\!\overset{\overset{\displaystyle Cl}{|}}{C}\!=\!CH\!-\!CH_2\!\right)_{\!n}$$

Chloroprene   Neoprene

Neoprene is a relatively expensive synthetic rubber, but is used when resistance to oils, chemicals, air, light and heat is required.

Natural Polyenes

Rubber is a polymer of isoprene and is a polyene with isolated double bonds. Nature uses many polyenes made up from isoprene units. Some of these have repeating conjugated diene systems, which makes them coloured. Common examples are **lycopene**, a red pigment in ripe tomatoes, and **carotenes**, present in many plants but first isolated from carrots. Carotene is converted in the human liver into vitamin A, which is necessary for the synthesis in the body of certain pigments essential to sight.

Conjugated Trienes

What happens to the chemistry when there are three double bonds all conjugated with one another as in hexa-1, 3, 5-triene, $CH_2\!=\!CH\!-\!CH\!=\!CH\!-\!CH\!=\!CH_2$? Reaction of one mol of this compound with one mol of bromine gives two products:

$$CH_2\!=\!CH\!-\!CH\!=\!CH\!-\!CH\!=\!CH_2 \xrightarrow{\;Br_2\;} \begin{cases} BrCH_2\!-\!CH\!=\!CH\!-\!CH\!=\!CH\!-\!CH_2Br \\ BrCH_2\!-\!CHBr\!-\!CH\!=\!CH\!-\!CH\!=\!CH_2 \end{cases}$$

In other words, as with buta-1, 3-diene, both 1, 2-addition and addition to the ends of the conjugated system may take place. The mechanism is

246

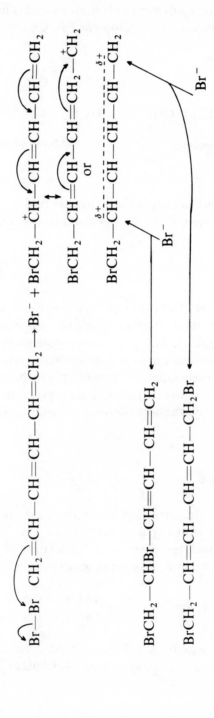

Trienes also take part in Diels–Alder reactions but only two of their conjugated double bonds are involved, for example,

Thus again **six** electrons are involved in the cyclo-addition reaction. If all three double bonds of the triene took part in the reaction, eight electrons would be involved, six from the triene and two from the other molecule; this does not happen.

Nor does a reaction such as

take place. This would involve four electrons, two from each reactant, in the reaction. The Diels–Alder reaction specifically requires six electrons from double bonds.

Cyclic Dienes and Trienes

Cyclohexene reacts like an acyclic alkene, e.g. it undergoes addition reactions with bromine or hydrogen chloride:

Similarly cyclohexa-1, 3-diene reacts like an acyclic conjugated diene:

1, 2-Addition   1, 4-Addition

It also takes part in Diels–Alder reactions, for example

When we come to the compound made up from six CH groups joined together in a ring, which could be written as

the resemblance to non-cyclic alkenes no longer obtains. This compound, which is called **benzene**, does not react with hydrogen chloride nor take part in Diels–Alder reactions. It only reacts with bromine in bright sunlight (or in ultraviolet light), but this is not an electrophilic addition reaction such as alkenes undergo. Chlorine also reacts only in bright sunlight (or ultraviolet light) and gives 1, 2, 3, 4, 5, 6-hexachlorocyclohexane, which is a very powerful insecticide.

If another of the standard chemical tests for the presence of unsaturation is applied, namely the decolourization of a solution of potassium permanganate (see Chapter 5), cyclohexene and cyclohexadiene react rapidly, but no reaction occurs with benzene. Perhaps even more strikingly if methylbenzene is treated with a solution of potassium permanganate, the saturated methyl group, which in most compounds is not readily oxidized, is converted into a carboxylic acid group, while the benzene ring remains unscathed:

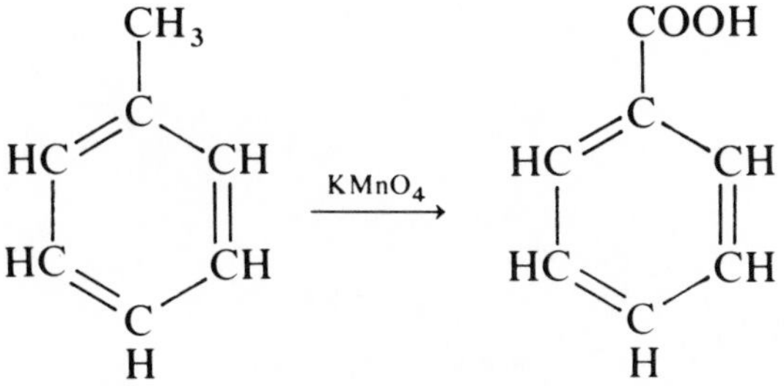

This cyclic molecule, benzene, C_6H_6, thus appears to be different from standard alkenes and polyenes, and to have a special stability and properties of its own. These properties, and the reasons for them, form the subject matter of the following chapters.

Questions

1. Give the mechanism of the reactions that take place between 1 mole of buta-1, 3-diene and 1 mole of hydrogen chloride.
2. What are the products of the reactions of hexa-2, 4-diene with (a) bromine and (b) but-3-en-2-one? Give both structural formulae and names of the products. Do you think that a similar reaction to that in (b) would take place between hexa-2, 5-diene and but-3-en-2-one?

24 BENZENE

Structure of Benzene

As has been seen in the previous chapter, benzene, C_6H_6, has chemical properties quite different from those of conventional alkenes, despite the fact that formally it might be considered as cyclohexatriene.

Its ^1H-n.m.r. and ^{13}C-n.m.r. spectra show that all the carbon atoms in benzene are equivalent to one another, and so are all the hydrogen atoms. This suggests a six-membered ring structure, with the atoms arranged as shown below (bonds are not included in this diagram):

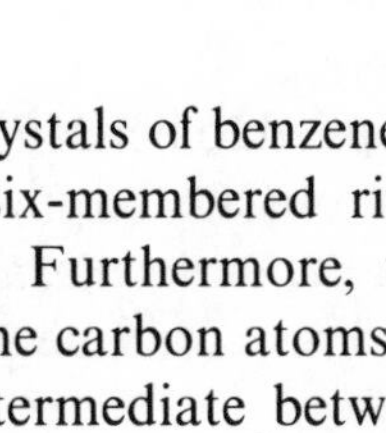

$$
\begin{array}{ccc}
 & H & \\
H & C & H \\
C & & C \\
 & & \\
C & & C \\
H & C & H \\
 & H &
\end{array}
$$

X-ray examination of crystals of benzene shows that it does indeed have this structure, the six-membered ring being made up of six equivalent C—H groups. Furthermore, the structure is that of a regular hexagon, and all the carbon atoms are at a distance of 1.40 Å apart. This distance is intermediate between the length of a single carbon–carbon single bond (~ 1.53 Å) and a simple carbon–carbon double bond (~ 1.34 Å).

To fulfil the normal valency of carbon it would be necessary to include alternate single and double bonds in the ring, as in

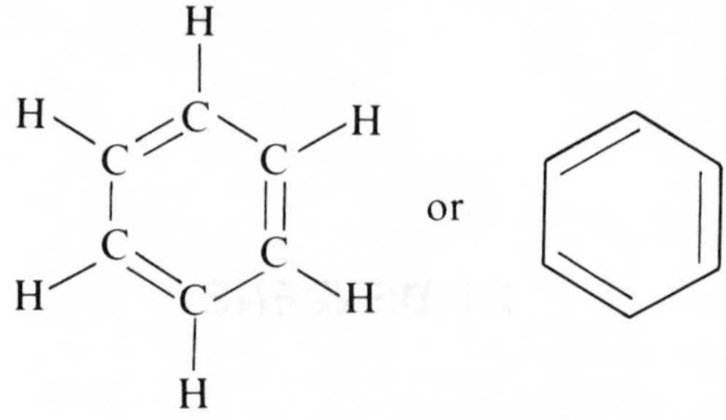

or

But this seems to be ruled out by the X-ray structure. It is possible that the X-ray evidence could be showing an average structure, which could be pictured as follows:

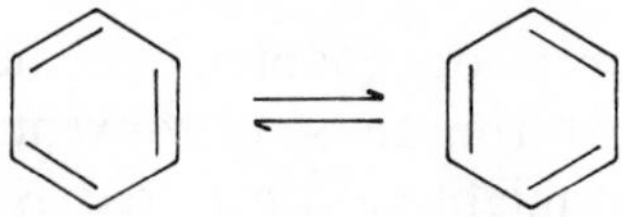

and that the double bonds and single bonds are exchanging positions too rapidly for either of the two alternate structures to be detected. Such interchange would proceed more slowly at lower temperatures but there is no evidence for anything other than an 'average' structure, even at extremely low temperatures.

The n.m.r. spectra also indicate that benzene is not a normal alkene, for the signals it provides appear at different positions from those expected for alkenes.*

In ethylene each carbon atom needs to use only three of its electrons for bonding to adjacent atoms (two to carbon and one to hydrogen atoms). Each carbon atom has a 'spare' electron; these electrons interact to form a double bond.

When this approach is applied to benzene, again only three electrons belonging to each carbon atom are required to hold the carbon and hydrogen atoms together in the molecule, and each carbon atom has a remaining 'spare' electron. This could be pictured as follows:

*In its ^{1}H-n.m.r. spectrum benzene provides a signal at δ 7.27; cf. protons attached to alkene double bonds at $\delta \sim 5.5$.

The six 'spare' electrons could interact with neighbours to form three double bonds, but all the spectroscopic and chemical evidence suggests the absence of normal double bonds.

Suppose that instead all six of these spare electrons interact with one another to form a group, a **sextet** of electrons.* This could be depicted as follows:

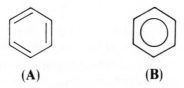

In this case there are no normal double bonds and the ring is completely symmetric. This representation would thus explain both the structure and symmetry of benzene, and its chemical properties. It is also reasonable that the lengths of the carbon–carbon bonds should be intermediate between the lengths of typical single and double bonds.

In benzene the six spare electrons are said to be **delocalized**, in contrast to the situation in simple alkenes in which the spare electrons are **localized** in the double bond.

Formulae for Benzene

Both formulae **A** and **B** are commonly used to represent benzene:

(A)   (B)

---

*This concept of the sextet of electrons and the accompanying formula was suggested by Robinson, at the time Professor of Chemistry in the University of St Andrews, in the early 1920s.

Formula **A** is known as the **Kekulé** formulation, after the chemist Kekulé, who played a major role in first elucidating the structure of benzene as a cyclic molecule in the 1860s. Formula **B** is called the inscribed circle formula or sometimes, more colloquially, as the poached egg formula. Both formulae have their advantages. **B** obviously shows the symmetry of the ring and lack of 'normal' double bonds, while A indicates at a glance that six 'extra' electrons are present. (Six electrons are needed to form the three 'extra' bonds.) On the whole, for reasons that are too complex to discuss here, formula **A** is probably the preferred representation, and will be used usually, but not exclusively, in this book. It is necessary to be conversant with both forms.

It is important to remember that all formulae are symbols or cartoons rather than exact pictures of molecules. Both formulae **A** and **B** are standard **symbols** used to represent the benzene ring.

Other Possible Delocalized Systems of Electrons

An important feature of the delocalized electronic structure of benzene is that **six** electrons are delocalized all round a six-membered ring.

In the preceding chapter the chemistry of non-cyclic conjugated trienes was mentioned. These trienes behave like normal conjugated alkenes and undergo the reactions of alkenes. So there is no delocalization of the six 'spare' electrons in such trienes, and all the chemical and physical evidence shows that a molecule such as hexa-1, 3, 5-triene has alternate single and double bonds, and is adequately symbolized as

$$CH_2{=}CH{-}CH{=}CH{-}CH{=}CH_2 \quad \text{or}$$

In discussions of cyclo-addition reactions in the previous chapter, it seemed that **cyclic** interaction of **six** electrons was important. Corresponding cyclo-addition reactions involving four or eight electrons do not take place. The same restriction appears to apply to cyclic polyenes. While benzene, with its six 'spare' electrons delocalized around a six-membered ring, is particularly stable and does not undergo typical reactions of alkenes, the corresponding four-membered-ring and eight-membered-ring analogues with alternate single and double bonds around the rings, shown below, have no such special properties, but behave as reactive alkenes:

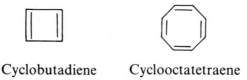

Cyclobutadiene   Cyclooctatetraene

Cyclooctatetraene reacts at once with bromine, or with a solution of permanganate. It is oxidized by air and slowly polymerizes. Cyclobutadiene is so reactive that it cannot even be isolated at room temperature.

Hence there is obviously something particularly important about the cyclic sextet of electrons which leads to the special stability and properties of benzene.

Anions or cations involving carbon, called **carbanions** or **carbocations**, can also be made more stable if they are associated with a cyclic delocalized system of six electrons.

Removal of a proton from cyclopentadiene gives an especially stable anion:

$$\overset{-H^+}{\longrightarrow}$$

This ion has six 'spare' electrons, which can interact around the ring. Two come from each of the double bonds, while removal of the proton has left another two electrons derived from the bond which previously had linked the hydrogen atom to the carbon atom. These six electrons can interact and be delocalized around the ring. At the same time the negative charge is also delocalized around the ring, and, as has been seen earlier in this book, delocalization of charge leads to greater stability of an ion. The cyclopentadienide ion is represented as follows:

The circle indicates that six electrons are delocalized around the ring, and the negative sign within the ring signifies that it too is delocalized. The ring is symmetric, being made up of five equivalent CH groups.

Sodium or potassium cyclopentadienides are made by the reaction of sodium or potassium with cyclopentadiene:

$$\text{(cyclopentadiene)} \quad \xrightarrow[\text{or K}]{\text{Na}} \quad \text{(cyclopentadienide}^-) \quad Na^+ \text{ or } K^+$$

Conversely, if a hydride ion, H^-, is removed from cycloheptatriene a stable delocalized cation results:

$$\text{(cycloheptatriene)} \quad \xrightarrow{-H^-} \quad \text{(cation)}$$

The carbon atom from which the hydride ion has been removed has no spare electrons, since the electrons which formed the carbon–hydrogen bond are associated with the hydrogen atom to from a hydride ion. The resultant carbocation has seven CH groups forming a seven-membered ring, and associated with this are six 'spare' electrons, derived from the three double bonds. These become delocalized around the ring, as does the positive charge, and the ion can be represented thus (it is known as a **tropylium** ion):

provides   Tropylium ion

The formation of these stable ions shows the importance of the cyclic interaction of **six** electrons. That the number of electrons, six, is important is shown by the facts that cyclopentadiene does not readily give a cation, which would have 'four' spare electrons, nor does cycloheptatriene readily provide an anion, which would have eight 'spare' electrons. Although in either of these latter cases some stabilization might be derived from delocalization of the charge around the ring, they are nonetheless not readily formed, since neither of them would have the essential total of six electrons.

Properties and Sources of Benzene

Benzene is a liquid at normal room temperature, and freezes to a solid at 5.5 °C. It is volatile, with a boiling point of 80 °C. As a hydrocarbon it is insoluble in water. It is dangerously inflammable and burns producing copious amounts of black soot. It is toxic both as a liquid and in its vapour form; it accumulates in the body, which is not able to dispose of it readily.

Benzene and its derivatives are very important industrial chemicals. Their use dates from the earliest days of the chemical industry, when they were obtained by distillation of coal and coal tar, and in the future this will undoubtedly again become a major source. At present the main source is from catalytic treatment of derivatives of oil and natural gas. Thus in the presence of a suitable catalyst, propane may be converted into benzene and hydrogen. Butane can be treated similarly to give dimethylbenzenes, **xylenes**, also industrially important compounds.

CH_3	OH	$CH=CH_2$	NH_2
Toluene	Phenol	Styrene	Aniline

Other common benzene derivatives of industrial importance are **toluene**, **phenol** and **styrene**. Benzene, toluene, xylene and phenol are all of great importance in the large-scale preparations of a large number of industrially important compounds, e.g. the starting materials for the preparation of nylon and of polyesters. **Polystyrene** is made by polymerization of styrene.

Nomenclature

Many simple benzene derivatives, e.g. toluene, xylene, phenol, styrene and aniline, have trivial names as their official names. Systematic nomenclature either treats compounds as derivatives of benzene in which a hydrogen atom is replaced by a substituent group, or makes use of the name of the group C_6H_5 (i.e. benzene less a hydrogen atom), which is **phenyl**.

Thus systematic names, not normally used, for compounds shown above are as follows:

$C_6H_5CH_3$	Toluene	Methylbenzene or phenylmethane
C_6H_5OH	Phenol	Hydroxybenzene
$C_6H_5NH_2$	Aniline	Aminobenzene or phenylamine
$C_6H_5CH{=}CH_2$	Styrene	Vinylbenzene or phenylethylene

Since the benzene ring is a regular hexagon, and all its carbon atoms are equivalent, it does not matter at which atom a single substituent group is drawn. Thus all the following formulae represent phenol:

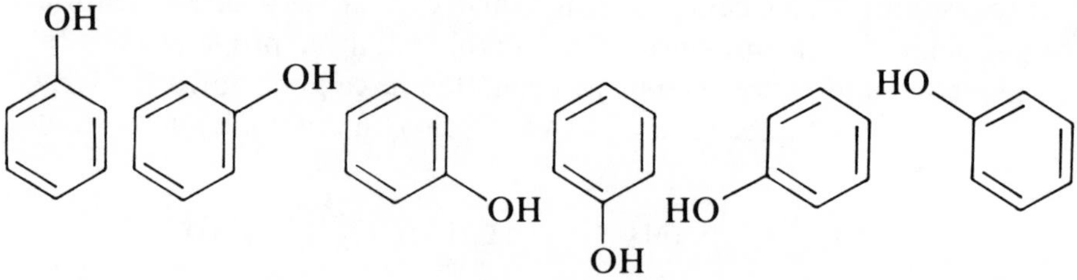

All of these formulae may be changed into one another by rotating them in the plane of the paper.

If, however, there are two or more substituent groups, the relative positions of these substituent groups must be indicated. For example, there are three isomers of dichlorobenzene, $C_6H_4Cl_2$:

(a) Two chlorine atoms attached to adjacent carbon atoms

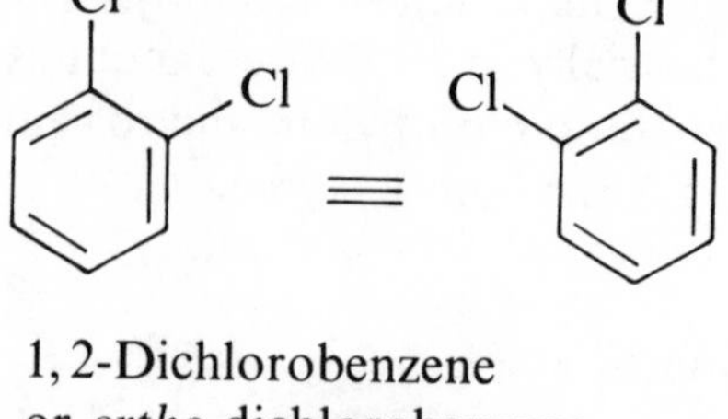

1, 2-Dichlorobenzene
or *ortho*-dichlorobenzene
or *o*-dichlorobenzene
(*o* is an abbreviation for *ortho*)

(b) Two chlorine
    atoms attached
    to carbon atoms
    separated by
    another carbon atom

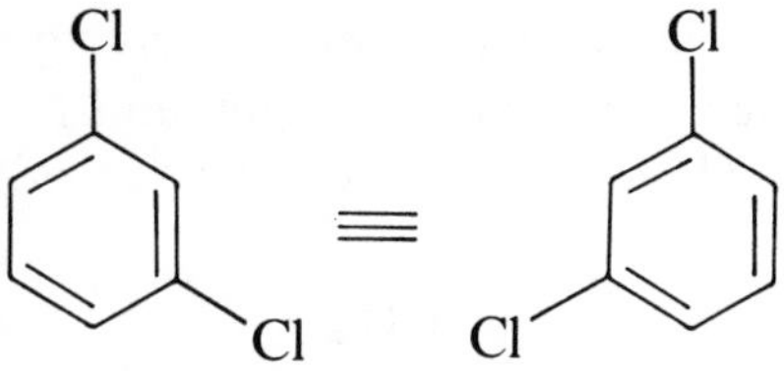

1, 3-Dichlorobenzene
or *meta*-dichlorobenzene
or *m*-dichlorobenzene

(c) Two chlorine
    atoms attached
    to carbon atoms
    separated by
    two carbon atoms

1, 4-Dichlorobenzene
or *para*-dichlorobenzene
or *p*-dichlorobenzene

Note that the two formulae given for 1, 2-dichlorobenzene are one and
the same compound:

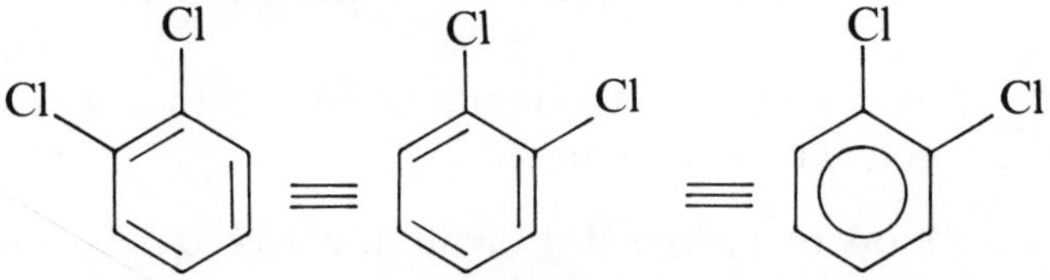

When the Kekulé formula for the benzene ring is used, as in formulae **C**
and **D**, if it were taken as a literal picture it might suggest that **C** and **D**
are different compounds with the two CCl groups linked by a double
bond in **C** and by a single bond in **D**. However, as emphasized before,
this representation is only a symbol; in fact there are no single or double
bonds and hence **C**≡**D**. This is at once evident when the inscribed
circle formula **E** is used.

The system of numbering is self-evident. Alternatively 1, 2-, 1, 3- and

1, 4- are replaced, respectively, by the terms *ortho-*, *meta-* and *para-*, or their abbreviations *o-*, *m-* and *p*.

Three dimethylbenzenes or xylenes exist:

o-Xylene   *m*-Xylene   *p*-Xylene

A similar system of nomenclature is used when two different substituents are present, for example

o-Bromochlorobenzene *m*-Chlorotoluene
or 2-bromochlorobenzene or 3-chlorotoluene

(In the numerical system the unnumbered substituent is at the 1 position; 1 is conventionally omitted.)

If there are more than two substituents attached to the benzene ring it is necessary to use the numerical system, for example

1, 3, 5-Trichlorobenzene 4-Bromo-2-chlorotoluene

2,4-Dichloro-3,5-dimethylphenol
(an ingredient of a common domestic disinfectant)

In general, compounds containing benzene rings are called, collectively, **benzene derivatives, aryl derivatives**, e.g. **aryl halides** and **arylamines**, or **aromatic compounds**. Hydrocarbons containing a benzene ring, such as benzene itself, toluene, the xylenes and styrene, are called **arenes**.

The chemistry of benzene and its derivatives will be considered in the next three chapters, under the following headings:

(a) chemical properties of the benzene ring itself,
(b) effects of substituents on the reactions of the benzene ring,
(c) effects of the benzene ring on reactions of substituent groups.

Questions

1. Draw the formulae of (a) p-chloroaniline, (b) m-bromoethylbenzene, (c) 4-bromo-2-chlorophenol, (d) 2,4,6-trinitrotoluene (nitro = NO_2) and (e) phenylethane.
2. Could you distinguish between the o-, m- and p-xylenes (o-, m- and p-dimethylbenzenes) by means of their ^{13}C-n.m.r. spectra?

25 REACTIONS OF BENZENE

Like alkenes, benzene is electron-rich; it has more valence electrons than are needed solely to keep all its constituent atoms linked together. Hence it should react with electrophiles. However, from its lack of reactivity towards bromine, or with hydrogen chloride, it is obviously less reactive than alkenes.

However, when benzene is added to very strong acids, e.g. cold concentrated sulphuric acid, n.m.r. spectra of the resultant solution show that it is to some extent protonated to give a cation $[C_6H_7]^+$. This ion is not very stable; hence the need for very strong acid. The n.m.r. spectrum further suggests that this ion has a structure as shown in the following equation:

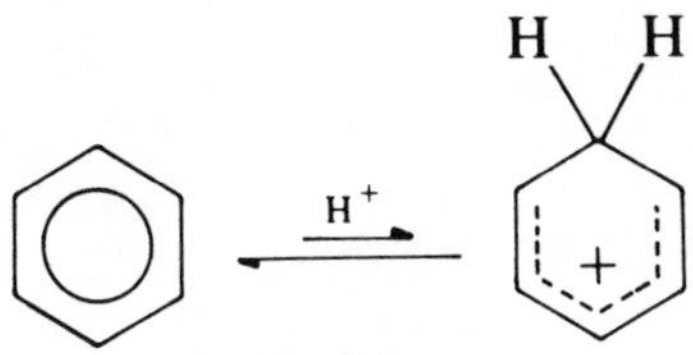

Two of the six electrons from the benzene ring are used to form a C—H bond. This leaves four electrons and a positive charge, which are delocalized over five atoms, as depicted in the above formula.

Formation of this cation can also be depicted making use of the Kekulé formula. Let us suppose that two of the electrons represented add to a proton in the same way that an alkene would:

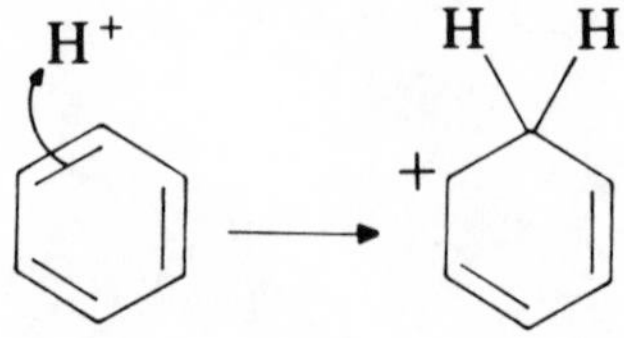

The remaining four electrons will interact with the positively charged carbon atom just as in acyclic dienes or trienes:

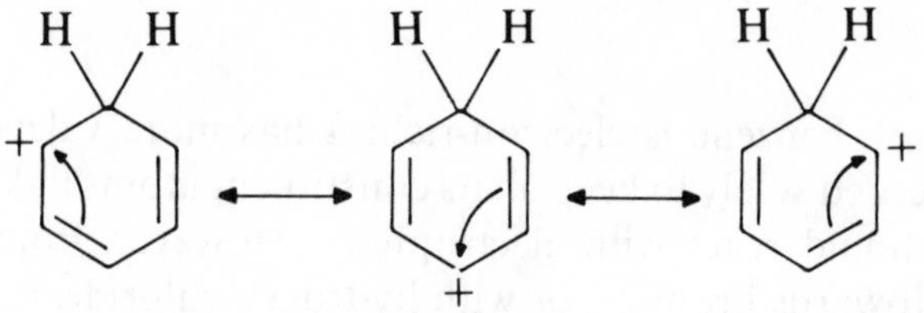

The mean of these formulae can be formulated as

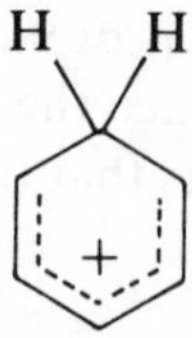

or better (but less simply) as

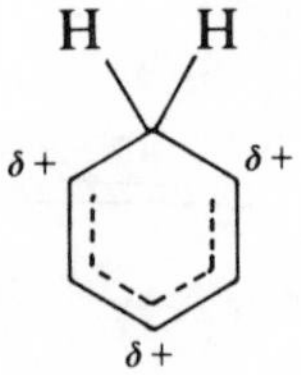

since the latter shows that the positive charge is especially located on the terminal and central carbon atoms of the delocalized system. This is found experimentally to be so.

This carbenium ion does not, however, react with the anion which is also present. Rather it just loses one of the hydrogen atoms of the CH_2 group and benzene is regenerated:

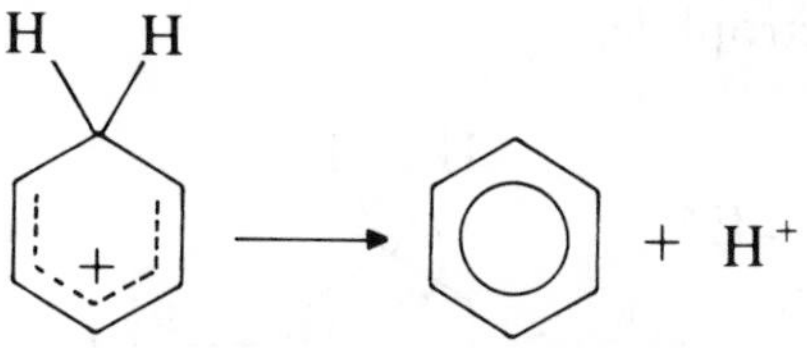

Thus, on dilution of the concentrated sulphuric acid solution with water the ion is lost:

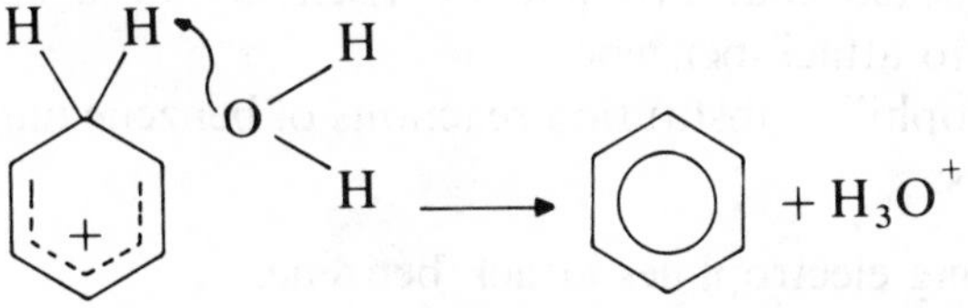

This very ready regeneration of benzene is a direct consequence of the great stability of the benzene system with its six delocalized electrons.

Deuteriation

If concentrated deuteriosulphuric acid is used instead of sulphuric acid a change may be effected, the end result being that a hydrogen atom is replaced by deuterium:

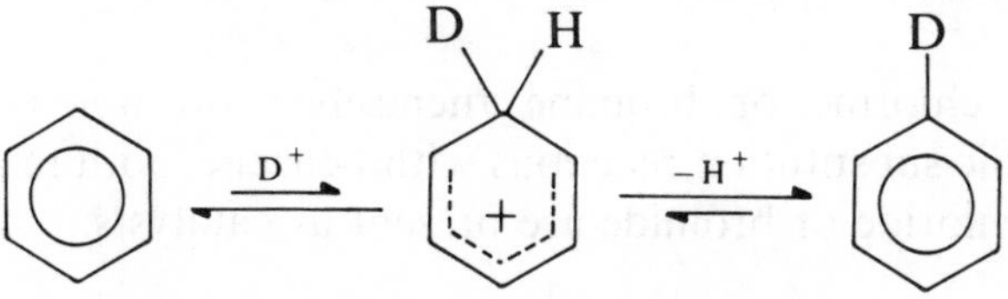

Thus a substitution reaction has taken place. Since the reagent is positively charged it is an electrophile, and the reaction is an **electrophilic substitution reaction**. Reaction takes place in two distinct steps: initial attack by the electrophile, followed by loss of a proton.

Other Electrophilic Substitution Reactions

Similar reactions may take place between benzene and other strong electrophiles. In general they may be summarized as follows, where E^+

stands for an electrophile:

It may be noted that bromine by itself is not a strong enough electrophile to attack benzene.

The electrophilic substitution reactions of benzene may be summed up as follows:

1. Only strong electrophiles attack benzene.
2. The net result of the reaction is **substitution**, and not addition, which takes place between alkenes and electrophiles.

These facts are consequences of the great stability of the benzene system with its six delocalized electrons. It takes a powerful reagent to attack and disrupt the electronic system, and in the second step it is restored, not destroyed.

Some of the commoner electrophilic substitution reactions of benzene will now be considered.

Chlorination and Bromination

Although chlorine or bromine themselves do not take part in electrophilic substitution reactions with benzene, both react if iron or iron(III) chloride or bromide are present as catalysts:

If iron is used as catalyst, it first reacts with some halogen to form iron halide.

The reaction is again substitution and the mechanism involves formation of a **bromonium** (or **chloronium**) cation, which is a much stronger electrophile than the halogens are. The mechanism is as follows:

$$Br\!-\!Br \quad FeBr_3 \longrightarrow Br^+ + [FeBr_4]^-$$

Chlorine reacts similarly in the presence of either iron(III) chloride or aluminium chloride.

It should be noted that reaction starts in the same way as an addition reaction of an alkene, but, instead of the intermediate adding an anion, it loses a proton:

$$CH_2\!=\!CH_2 \quad Cl\!-\!Cl \longrightarrow \overset{+}{C}H_2\!-\!CH_2Cl \longrightarrow ClCH_2\!-\!CH_2Cl$$
$$+ \ Cl^-$$

This again reflects the great stability of the benzene system.

Nitration

When benzene is treated with a mixture of concentrated nitric and sulphuric acids, **nitration** takes place.

Nitrobenzene

Sulphonation

When benzene is heated with concentrated sulphuric acid, or treated with fuming sulphuric acid at room temperature, **sulphonation** occurs and a **sulphonic acid** is formed:

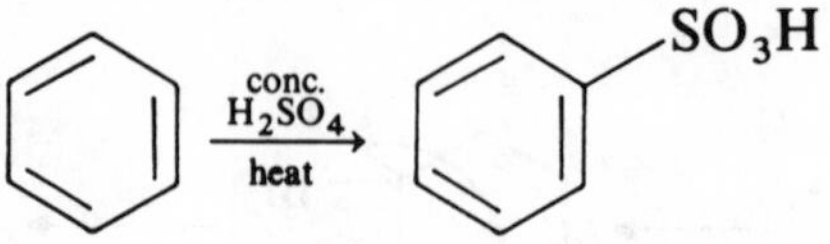

Benzenesulphonic acid

Sulphonic acids are strong acids. They form amides, called **sulphonamides**, for example $C_6H_5SO_2NR_2$. Some more complex sulphonamides have been widely used in medicine as antibacterial agents.

Acylation and Alkylation. Friedel–Crafts Reactions

Benzene reacts with acid chlorides or with alkyl halides in the presence of aluminium chloride to form acyl or alkyl benzenes:

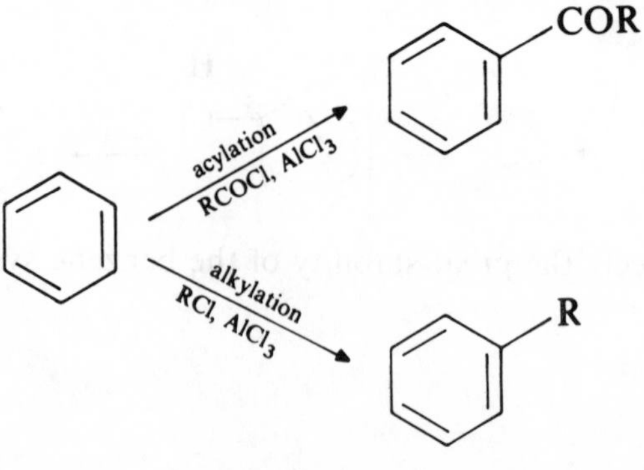

Acylation is commonly an effective reaction and provides **aryl ketones**, but alkylation does not proceed so satisfactorily except in simple cases.

These reactions are jointly called **Friedel–Crafts Reactions**, after their discoverers.

Mechanisms of Reactions

The precise mechanisms of the above reactions are complex and the following are summarized versions.

Sulphuric acid is a stronger acid than nitric acid and in a mixture of the two the nitric acid acts as a base and loses $[HO]^-$ to form the

nitronium ion $[NO_2]^+$. The detailed equilibrium leading to its formation is

$$HNO_3 + 2H_2SO_4 \rightleftharpoons H_3O^+ + NO_2^+ + 2HSO_2^-$$

The essence of its formation can be simplified to

$$HO\!\!\underset{\displaystyle H^+}{\diagup}\!\!NO_2 \rightarrow H_2O + NO_2^+$$

The nitronium ion is a strong electrophile which attacks the benzene ring.

Sulphonation is more complex but in essence involves protonation of sulphuric acid molecules to form $[H_3SO_4]^+$ which acts as an electrophile:

$$HOSO_2OH \xrightleftharpoons{H^+} H_2\overset{+}{O}SO_2OH \longrightarrow$$

$$C_6H_5SO_3H$$

In Friedel-Crafts reactions the acyl or alkyl chlorides react with aluminium chloride to give carbenium ions, which then react with the benzene ring:

$$RCOCl + AlCl_3 \rightleftharpoons [RCO]^+[AlCl_4]^- \xrightarrow{C_6H_6} C_6H_5COR$$

$$RCl + AlCl_3 \rightleftharpoons R^+[AlCl_4]^- \xrightarrow{C_6H_6} C_6H_5R$$

Reduction of Benzene

Like alkenes, benzene can be reduced by hydrogen in the presence of a catalyst, but reaction is more difficult in the case of benzene. The product is cyclohexane:

Because of the difference of reactivity it is possible in the case of molecules containing both a benzene ring and on alkene group to reduce the latter preferentially:

Using more vigorous conditions the benzene rings can be reduced as well.

Also because of this difference in reactivity between benzene and alkenes it is very difficult to reduce benzene partially, to provide cyclohexadiene. As soon as any of the latter is formed it reacts more rapidly than benzene, giving cyclohexene and cyclohexane, and successfully competes with benzene for the hydrogen.

Questions

1. How may benzene be converted into (a) nitrobenzene, (b) benzene-sulphonic acid, (c) acetylbenzene (methylphenylketone and (d) deuteriobenzene?
2. Give the mechanism of the reactions involved in (a) bromination and (b) acylation of benzene.

26 EFFECTS OF SUBSTITUENTS ON REACTIONS OF THE BENZENE RING

The effect of substituents on the reactions of the benzene ring is particularly evident and important when it is required to introduce a second substituent into a monosubstituted benzene derivative, as in

(The position of Y, *o*, *m* or *p*, is undefined.)

Perhaps not surprisingly, some substituent groups appear to make the benzene ring more reactive towards electrophiles; other substituents make it less reactive.

Thus **activating groups** include $-CH_3$, $-OCH_3$, $-N(CH_3)_2$, $-OH$ (i.e. alkyl, ether, amine and hydroxy groups).

Deactivating groups include $-NO_2$, $-COR$, $-CHO$, $-COOH$, $-SO_3H$ (i.e. nitro, acyl, aldehyde, carboxylic acid and sulphonic acid groups).

Since the reactions take place between the benzene ring and electrophiles it might be expected that substituent groups that can increase the electron density in the ring should assist reaction, while substituent groups that tend to lower the electron density in the ring should make it less reactive. This is indeed the case.

The bond linking a methyl group is polarized towards the benzene ring, $C_6H_5 \leftarrow CH_3$, and thus can increase the electron density in the ring. Oxygen or nitrogen atoms next to the benzene ring, as in ethers, phenols and amines, can interact with the ring by means of their lone pairs of electrons, $C_6H_5 \overset{\frown}{\ddot{O}}CH_3$, $C_6H_5 \overset{\frown}{\ddot{O}}H$, $C_6H_5 \overset{\frown}{\ddot{N}}H_2$. Hence they too can supply electrons to the benzene ring. All these groups are described as **electron-donating groups**.

In the case of deactivating groups, all of them have arrangements of atoms that tend to draw electrons away from the benzene ring:

$$C_6H_5-N\overset{\displaystyle O}{\underset{\displaystyle O}{}} \qquad C_6H_5-C\overset{\displaystyle O}{\underset{\displaystyle R}{}} \qquad C_6H_5-C\overset{\displaystyle O}{\underset{\displaystyle H}{}}$$

$$C_6H_5-C\overset{\displaystyle O}{\underset{\displaystyle OH}{}} \qquad C_6H_5-\overset{\displaystyle O}{\underset{\displaystyle O}{S}}-OH$$

These groups are described as **electron-withdrawing groups**.

It is possible to provide simple qualitative examples of these effects, as follows:

1.

To substitute a second nitro group into a benzene ring requires stronger acid than to introduce the first nitro group.

2.

Whereas benzene does not react with bromine except in the presence of a catalyst such as iron(III) bromide, phenol, with its electron-donating hydroxy group, reacts readily with an aqueous solution of bromine, without catalyst, to give a tribromophenol.

3.

Picric acid

Whereas benzene needs a mixture of concentrated nitric acid and concentrated sulphuric acid for its conversion into nitrobenzene, phenol requires only dilute aqueous nitric acid to convert it into nitrophenols. The same conditions which convert benzene into nitrobenzene lead to the conversion of phenol into the trinitro-phenol, which is called picric acid.

4. It is easier to make trinitrotoluene (TNT) from toluene than it is to make trinitrobenzene from benzene. This is one of the reasons for the wide use of TNT as an explosive.

Sites of Attack in Reactions of Substituted Benzene Derivatives

In the examples of electrophilic substitution reactions of different substituted benzenes which are mentioned in the preceding para-graphs, in each case the site of the second substituent is shown as a specific position, *o*, *m* or *p*. This represents the major product; in each case certain sites are markedly preferred to others.

On purely statistical grounds it could be expected that substitution reactions of a compound C_6H_5X might give products C_6H_4XY in the ratios $o:m:p = 2:2:1$, since this reflects the numbers of sites available for reaction.

In fact it is found that if the first substituent is an electron-donating group, further substitution takes place mostly (although not entirely) at the *o* and *p* positions, while if the first substituent is an electron-withdrawing group, further substitution takes place mainly (but again not entirely) at the *m* position.

The position of substitution is controlled by the first step of the reaction:

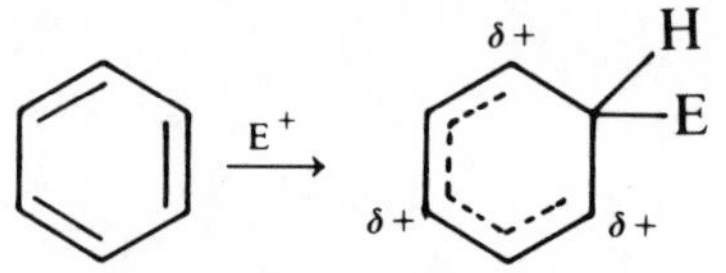

In the following discussion an important factor is the location of positive charge in the intermediate thus formed, as shown in the above formula.

Consider the effect of an electron-donating group X, such as NH_2 or OCH_3:

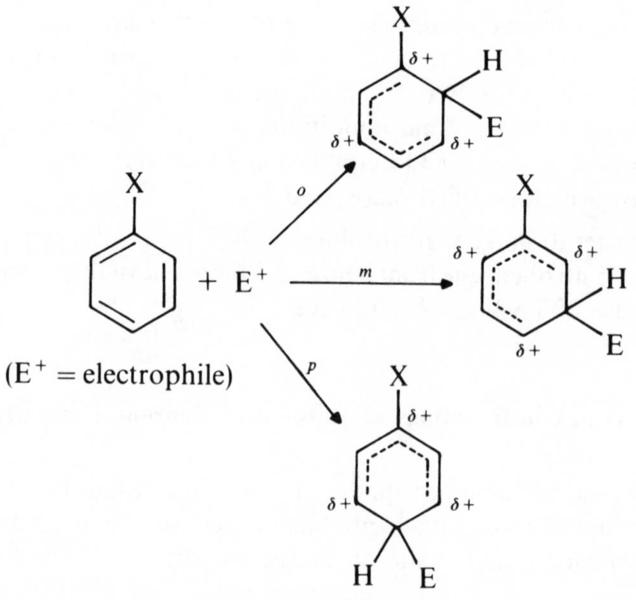

$(E^+ = \text{electrophile})$

In the case of the intermediates formed from attack at the *o* and *p* positions, the electron-donating substituent X is attached to one of the carbon atoms which bears a partial positive charge. The electron-donating character of X helps to neutralize this positive charge. In consequence these intermediates are more stable than that formed by attack of the electrophile at the *m* position, in which the electron-donating group is not directly attached to one of the partially positively charged carbon atoms.

This situation can be presented in terms of energy profiles:

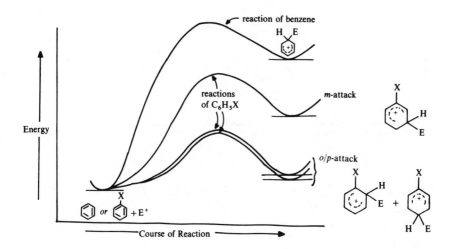

The above diagram shows the energy profiles for the reaction of some electrophile, E^+, at the *o*, *m* and *p* positions of a substituted benzene C_6H_5X, in which X is an electron-donating group, and on benzene itself.

A number of points are illustrated in this energy profile:

1. The activation energies of these reactions are related to the relative energies of the intermediates formed. In consequence, the more stable the intermediate, the lower the activation energy for its formation and the more readily and rapidly reaction proceeds.

2. Reaction at any site in C_6H_5X is faster than the same reaction involving benzene, because of the overall effect of the electron-donating group X.

3. Reaction at the o and p positions is particularly favoured, as discussed previously, and hence is the predominant reaction.

If an electron-withdrawing group Y is attached to the benzene ring in a compound C_6H_5Y, the opposite situation arises:

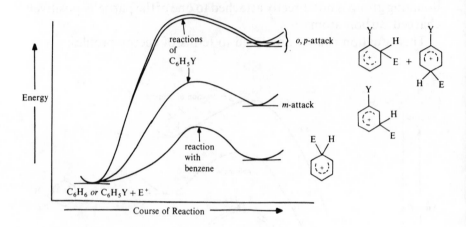

In this case attack at any site in the ring in C_6H_5Y is more difficult (has a greater activation energy) than for the same reaction involving benzene, and attack at the o and p positions is particularly disfavoured.

Consider the reactions involving a typical example of an electron-withdrawing group, CHO. This is deactivating because the carbonyl group is polarized away from the benzene ring, resulting in a partially positively charged atom being located next to the benzene ring, viz.

Formation of the initial intermediates in electrophilic substitution reactions at different positions in the ring are as follows:

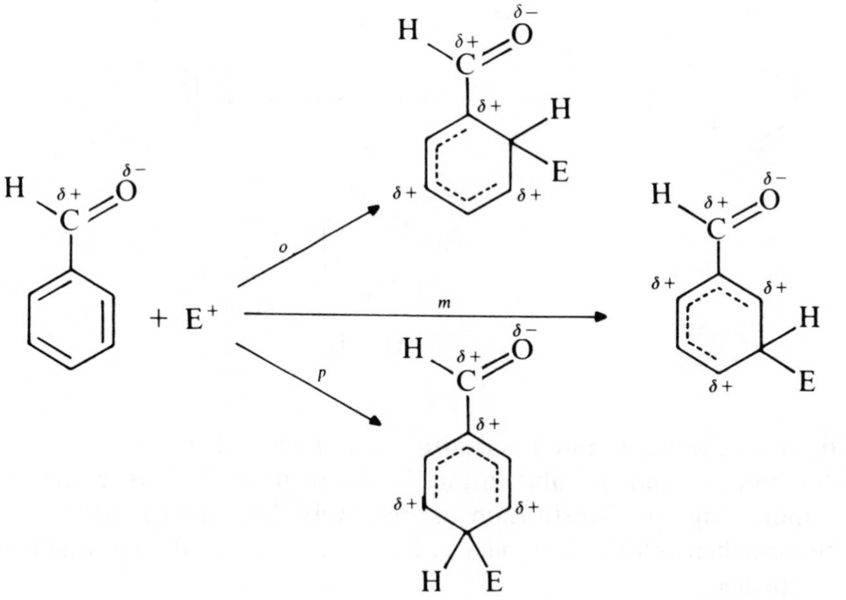

Reaction at the *o* and *p* positions results in some positive charge being developed on a ring atom to which the electron-withdrawing substituent is attached. There is repulsive interaction between the adjacent positive charges, which increases the energy. Attack at the *m* position does not give rise to adjacent positively charged atoms and hence requires less energy. Therefore reaction proceeds mainly at the *m* position. Substitution at all positions proceeds less readily than for benzene because of the overall effect of the electron-withdrawing group Y.

Summing up these considerations, it may be noted that electrophilic attack at substituted benzenes, C_6H_5Z, produces two patterns of charge distribution in the intermediate product. In both *o* and *p* substitution a partial positive charge is produced on the carbon atom to which Z is attached; in *m* substitution this is not the case:

In consequence, o and p substitution is favoured if Z is electron-donating, o and p substitution is disfavoured if Z is electron-withdrawing. m Substitution is relatively less influenced by Z because there is little, if any, charge located on the ring atom to which it is attached.

This may be summarized in tabular form as follows:

Nature of substituent	Effect of reaction at o and p positions	Effect of reaction at m position	Major substitution products
Electron-donating	Intermediate stabilized; reaction favoured	Small	o, p
Electron-withdrawing	Intermediate destabilized; reaction disfavoured	Small	m

Influence on the Order of Introducing Substituents into a Benzene Ring

If two substituent groups need to be introduced into a benzene ring, the order in which this is done is important if particular isomers are required. Thus in the making of chloronitrobenzenes, the nitro group is introduced first if the *m*-isomer is desired; if the reaction sequence is reversed the *o*- and *p*-isomers are obtained as the main products:

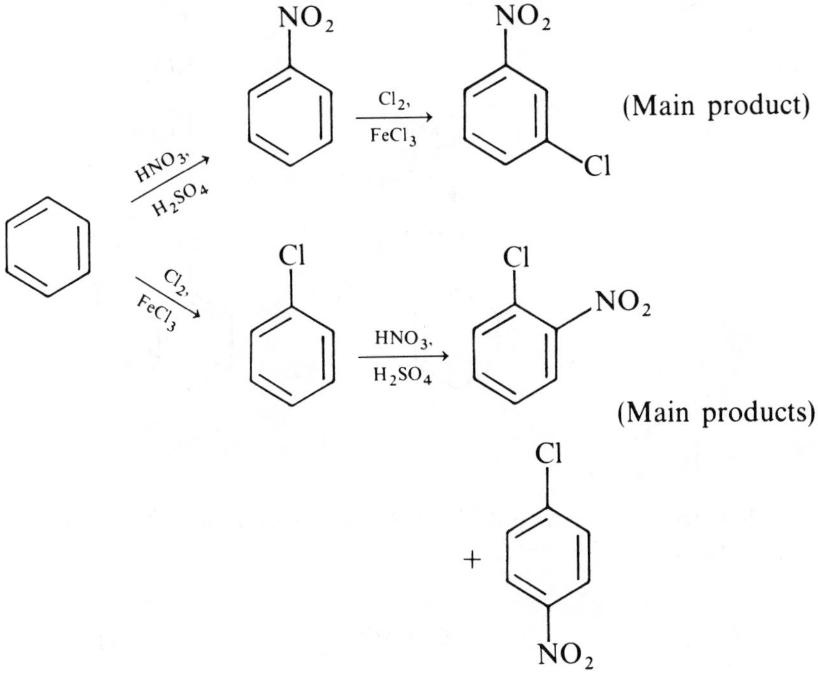

It is sometimes necessary to use indirect methods in order to obtain specific isomers. For example, nitration of benzoic acid, C_6H_5COOH, leads to the formation of *m*-nitrobenzoic acid:

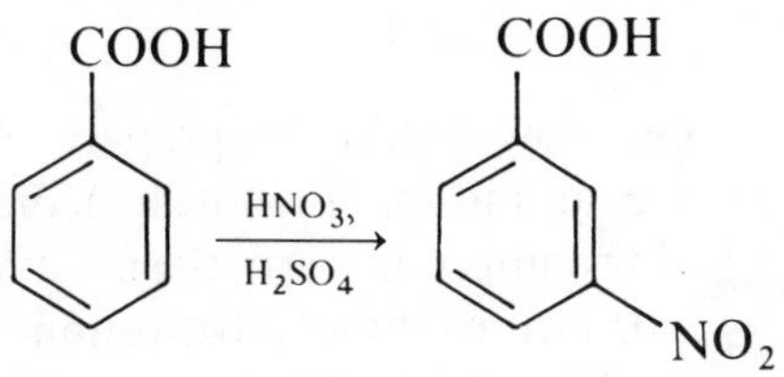

In order to obtain the *o*- and *p*-isomers, it is necessary to nitrate toluene and then to oxidize the *o*- and *p*-nitrotoluenes:

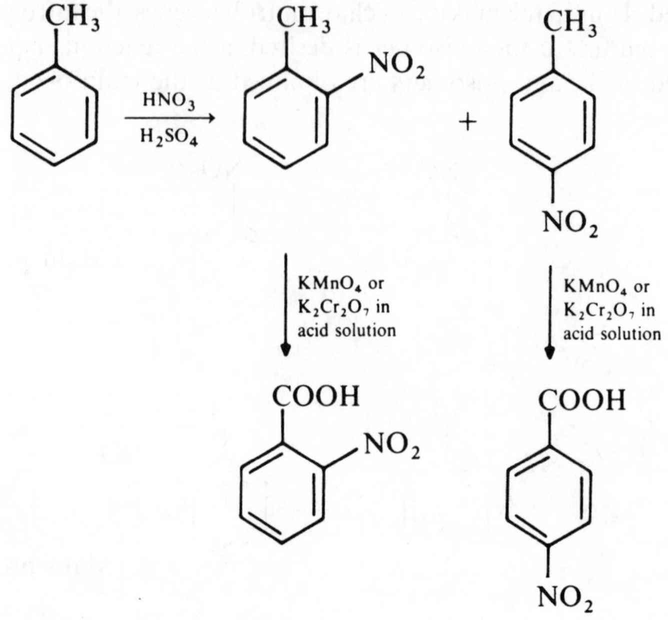

m-Nitrobenzoic acid could be obtained from toluene by oxidizing it first to benzoic acid:

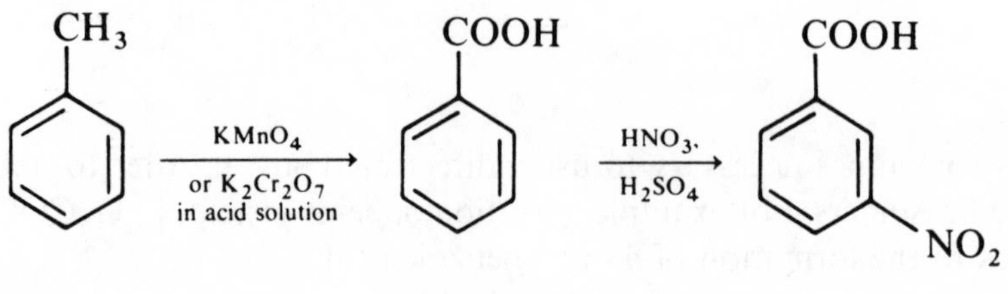

Questions

1. Starting from toluene, how would you prepare (a) *p*-chlorotoluene, (b) *p*-chlorobenzoic acid and (c) *m*-chlorobenzoic acid?
2. Which of the following compounds is (i) the most reactive and (ii) the least reactive towards electrophilic substitution?
   (a) benzene, (b) benzaldehyde, (c) methoxybenzene.
   Explain briefly.
3. How many signals appear in the ^{13}C-n.m.r. spectra of the products obtained when

(a) nitrobenzene is treated with a mixture of fuming nitric acid and concentrated sulphuric acid,

(b) phenol is treated with one mol of bromine

(c) phenol is treated with three mols of bromine?

Write equations for the reactions and indicate on the formulae of the products the carbon atoms associated with different signals in their ^{13}C-n.m.r. spectra.

27 EFFECTS OF THE BENZENE RING ON SUBSTITUENT GROUPS

Effects on Carbonyl Groups

Carbonyl groups attached to a benzene ring are little affected by it in their chemical properties. Thus arylaldehydes, such as **benzaldehyde**, C_6H_5CHO, aryl ketones, such as **acetophenone**, $CH_3COC_6H_5$, and **benzophenone**, $C_6H_5COC_6H_5$, and aryl carboxylic acids, such as **benzoic acid**, C_6H_5COOH, closely resemble their alkyl analogues, although the reactivity of the carbonyl group of the aldehydes and ketones is somewhat lower in the case of aryl derivatives.

Many aryl carboxylic acids are insoluble in cold water, but, because they react with aqueous sodium hydroxide to form soluble sodium salts, dissolve in this solution. Neutralization of the resultant solution with strong acid causes precipitation of the carboxylic acid:

$$C_6H_5COOH \underset{\text{HCl}}{\overset{\text{NaOH}}{\rightleftharpoons}} C_6H_5CO_2^- Na^+$$

| Insoluble in cold water | Soluble in cold water |

Some other groups are much more influenced by the presence of a phenyl group. In this chapter, the effects of the benzene ring on substituent alkyl groups, halogen atoms, hydroxy groups and amino groups will be considered.

Alkylbenzenes

Mention has been made previously of the relatively ready oxidation of substituent methyl groups to give carboxyl groups. for example

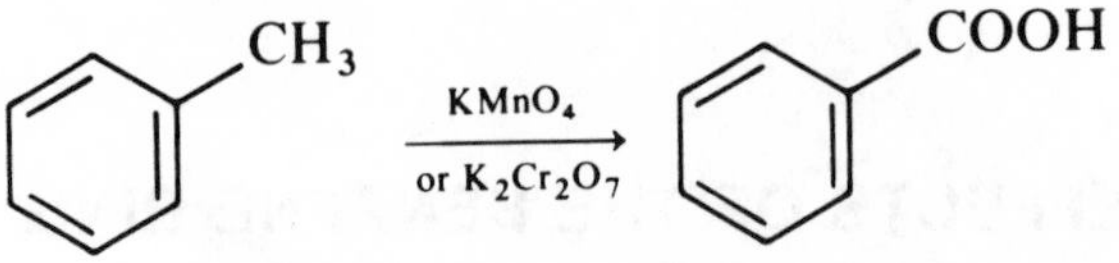

Benzoic acid

All substituent alkyl groups may be oxidized similarly and are converted into carboxyl groups, for example

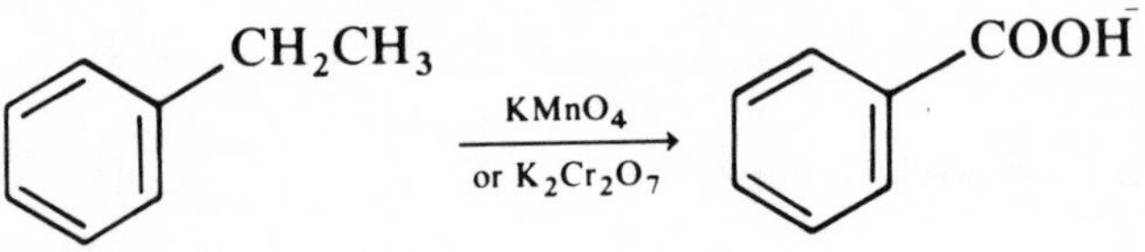

A very important industrial process is the oxidation of *p*-xylene to give terephthalic acid:

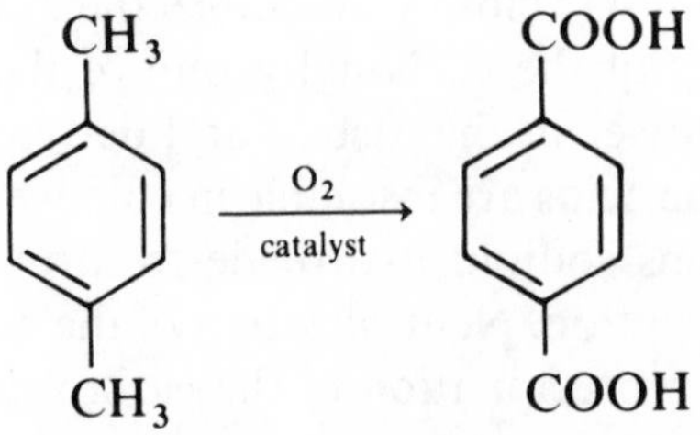

Terephthalic acid

As has been seen in Chapters 17 and 19, terephthalic acid is extremely important for its use in the preparation of polyesters such as 'Terylene' and 'Dacron', and polyamides such as 'Kevlar' and Twaron'. In the preparation of these polyamides the other reactant is also a benzene derivative, *p*-diaminobenzene.

Another characteristic reaction of alkylbenzenes is that they are readily halogenated at the carbon atom next to the benzene ring. The reaction is catalysed by light:

$$C_6H_5CH_3 \xrightarrow[hv]{Cl_2} C_6H_5CH_2Cl \rightarrow C_6H_5CHCl_2 \rightarrow C_6H_5CCl_3$$

$$C_6H_5CH_2CH_3 \xrightarrow[hv]{Cl_2} C_6H_5CHClCH_3 \rightarrow C_6H_5CCl_2CH_3$$

As in the chlorination of alkanes, this is a reaction involving radicals, and reaction is particularly easy because the intermediate radical, for example $C_6H_5\dot{C}H_2$ or $C_6H_5\dot{C}HCH_3$, is stabilized by the adjacent benzene ring.

Alkylbenzenes can thus be chlorinated either in the ring or on the substituent alkyl group, depending on the reaction conditions employed:

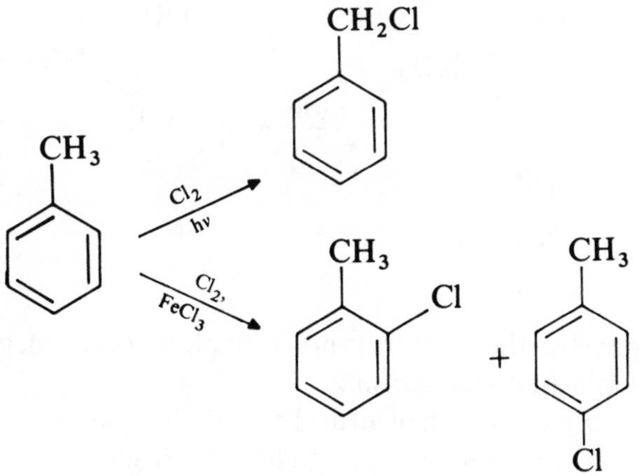

Aryl Halides

Unlike alkyl halides, most aryl halides are unreactive towards nucleophiles. Two factors contribute. Nucleophiles react at electron-poor sites, whereas the benzene ring is electron-rich. The bulky benzene group hinders the rearward approach of the reagent which is a requisite of S_N2 reactions. Thus, under the conditions used for reactions of alkyl halides with hydroxides to give alcohols, no reaction takes place with aryl halides.

In industrial conditions, where high-temperature and high-pressure conditions are more readily available, the reaction can be made to proceed, for example

$$C_6H_5Cl \xrightarrow[\substack{\text{high temperature,} \\ \text{high pressure}}]{\text{NaOH}} C_6H_5OH$$

This is used as one method for the commercial preparation of phenol. Reaction does not involve direct substitution of chlorine by hydroxide, but a different mechanism.

Other nucleophiles similarly do not react with aryl halides under the conditions that are used for their reactions with alkyl halides.

Aryl halides which also have strong electron-withdrawing groups attached to the benzene ring react with hydroxides to give phenols, for example

However, here too the reaction is not a simple S_N reaction, but involves a more complicated mechanism.

The only simple reaction of an aryl halide which can easily be done in the laboratory is with magnesium to form a Grignard reagent or with lithium to give a lithium derivative. Chlorobenzene is much less reactive than bromobenzene in these reactions.

Thus aryl carboxylic acids can be made from aryl halides.

Compounds with halogen atoms attached to substituent alkyl groups react similarly to alkyl halides, for example

$$C_6H_5CH_2Cl \underset{\overset{KCN}{\searrow}}{\overset{\overset{NaOH}{\nearrow}}{}} \begin{array}{l} C_6H_5CH_2OH \\ \\ C_6H_5CH_2CN \end{array}$$

Phenols

Phenols are hydroxy derivatives of benzene. As chlorobenzene is not readily converted into phenol, likewise phenol is not readily converted into chlorobenzene. Phenol does not react with carboxylic acids to form esters, although esters can be obtained from reactions with acid anhydrides or acid chlorides, for example

$$C_6H_5OH + CH_3COCl \longrightarrow CH_3COOC_6H_5$$
Phenyl acetate

The most interesting difference between phenols and alcohols is that phenols are weakly acidic. As discussed in Chapter 15, carboxylic acids are much more acidic than alcohols because the anions they provide are stabilized by delocalization of the negative charge between two equivalent oxygen atoms:

$$R-O-H \rightleftharpoons RO^- + H^+$$

In the case of phenols the negative charge in the anion can be delocalized to some extent into the benzene ring; this results in the **phenoxide** anion being more stable than an alkoxide anion.

Using the Kekulé type of formula, this delocalization of the negative charge can be pictured as follows:

In other words, the negative charge is shared between the oxygen atom and three sites, *o* and *p* to the hydroxy group, in the ring

This delocalization is not as effective as that in the carboxylate anion, wherein the negative charge is shared between two equivalent oxygen atoms, so the stabilities of the anions are in the order

carboxylate > phenoxide > alkoxide

Thus the acidities are in the order

carboxylic acids > phenols > alcohols

   Carboxylic acids are strong enough acids to form salts when treated with either sodium hydroxide or sodium carbonate. Phenols form salts with sodium hydroxide but not with sodium carbonate, while alcohols do not give salts with either of these reagents. In consequence carboxylic acids commonly dissolve in aqueous solutions of sodium hydroxide or sodium carbonate, whereas phenols dissolve in aqueous sodium hydroxide but not in aqueous sodium carbonate. Phenols which have electron-withdrawing substituents attached to the benzene ring, such as picric acid, 2, 4, 6-trinitrophenol, are much stronger acids. (For an X-ray structure of the picrate anion see the Appendix.)

Arylamines

Arylamines are made by the reduction of nitro compounds. Thus aniline is obtained from nitrobenzene (and hence from benzene):

As the benzene ring affects the acidity of phenols, so it affects the basicity of arylamines. In the equilibrium

$$C_6H_5NH_2 + H^+ \rightleftharpoons C_6H_5\overset{+}{N}H_3$$

the equilibrium is much more on the side of the amine than is the case for alkylamines, and arylamines are much weaker bases. In this case it is the stabilization of the amine by electronic interaction of the amino group with the benzene ring that increases the stability of the amine:

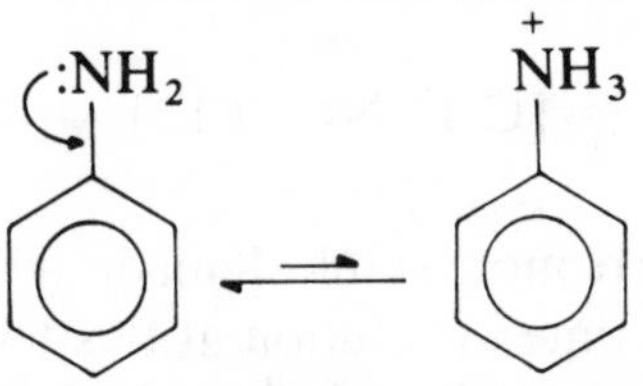

In forming the anilinium cation the lone pair of electrons on the nitrogen atom are utilized and thus can no longer interact with the ring. This tilts the equilibrium towards the amine. No such considerations apply in the case of alkylamines and they are consequently stronger bases.

Arylamines do, however, form salts with strong acids:

$$C_6H_5NH_2 + HCl \rightleftharpoons C_6H_5\overset{+}{N}H_3 \quad Cl^-$$
$$\text{Anilinium chloride}$$

In consequence aniline and many other arylamines are soluble in aqueous hydrochloric acid. Addition of base regenerates aniline, which is only slightly soluble in water and hence separates out:

$$C_6H_5\overset{+}{N}H_3 \quad Cl^- \xrightarrow{\text{NaOH}} C_6H_5NH_2$$

Arylamines, like alkyl amines, react with acid chlorides or acid anhydrides to form amides, for example

$$C_6H_5NH_2 + (CH_3CO)_2O \rightarrow C_6H_5NHCOCH_3$$
$$\text{Acetanilide}$$

Diazonium Salts

An important reaction of primary arylamines is with nitrous acid, generated from a mixture of hydrochloric acid (or other acid) and sodium nitrite. The product is called a **diazonium salt**:

$$C_6H_5NH_2 + NaNO_2 + HCl \longrightarrow C_6H_5N_2^+ \quad Cl^-$$
$$\text{Benzene diazonium}$$
$$\text{chloride}$$

All primary amines, alkyl as well as aryl, react similarly, but alkyl diazonium salts decompose immediately with evolution of nitrogen and, if the reaction has been carried out in aqueous solution, the main product is commonly an alcohol, for example

$$C_2H_5NH_2 \xrightarrow[\text{HCl,H}_2\text{O}]{\text{NaNO}_2,} [C_2H_5N_2^+ \quad Cl^-] \longrightarrow C_2H_5OH + N_2$$

Aryl diazonium salts are more stable. If made at $0\,^\circ C$ or just below they can be kept for some time in solution at this temperature, but above $0\,^\circ C$ they also decompose with evolution of nitrogen. Many of the solid salts can decompose explosively. Aryl diazonium salts are important reagents; they are usually made and used in solution at $\sim 0\,^\circ C$.

One important type of reaction that diazonium salts undergo is nucleophilic substitution reactions, which proceed by an S_N1 mechanism. Nitrogen is an extremely good leaving group, and reaction is irreversible. Hence aryl diazonium salts play a similar type of role to that of alkyl halides, and are of similar importance as synthetic reagents.

When diazonium salts are heated in aqueous solution the following reaction takes place:

$$C_6H_5N_2^+ \xrightarrow{\text{heat}} C_6H_5^+ + N_2 \longrightarrow C_6H_5\overset{+}{O}H_2 \longrightarrow C_6H_5OH$$
$$Cl^- \qquad\qquad \overset{\displaystyle O}{\underset{\displaystyle H \quad\; H}{}}$$

If other nucleophiles are present, similar reactions may occur, for example

$$C_6H_5N_2^+ \quad Cl^- \xrightarrow{\text{KI}} C_6H_5I + N_2$$

Often these reactions proceed less satisfactorily, and an alternative method uses copper(I) salts as reagents; these reactions involve a modified reaction mechanism. Examples include:

$$C_6H_5N_2^+ \quad Cl^-$$

with branches:
- $\xrightarrow[\text{heat}]{CuCl} C_6H_5Cl$
- $\xrightarrow[\text{heat}]{CuBr} C_6H_5Br$
- $\xrightarrow[\text{heat}]{CuCN} C_6H_5CN$

The latter reaction provides an alternative way of obtaining carboxylic acids from arenes, since aryl cyanides can be hydrolysed in the same way as their alkyl analogues:

$$C_6H_6 \xrightarrow[H_2SO_4]{HNO_3,} C_6H_5NO_2 \xrightarrow[HCl]{Sn,} C_6H_5NH_2$$

$$\downarrow \text{NaNO}_2, \text{HCl}$$

$$C_6H_5COOH \xleftarrow[H_2O]{H^+,} C_6H_5CN \xleftarrow{CuCN} C_6H_5N_2^+ \quad Cl^-$$

Diazonium salts can be reduced to arylhydrazines:

$$C_6H_5N_2^+ \quad Cl^- \xrightarrow[HCl]{SnCl_2,} C_6H_5NHNH_2$$

Phenylhydrazine

Since diazonium salts are positively charged they are electrophiles, although not very reactive ones. They will, however, react with compounds having activated benzene rings such as phenols:

The product is an **azo compound**. Simple azo compounds are yellow to red in colour, and many are used as dyestuffs.

Diazonium salts react similarly with tertiary arylamines, for example

Reactions with phenols and tertiary arylamines normally take place at the *p*-position, but if this position is already substituted, reaction takes place at the *o* position:

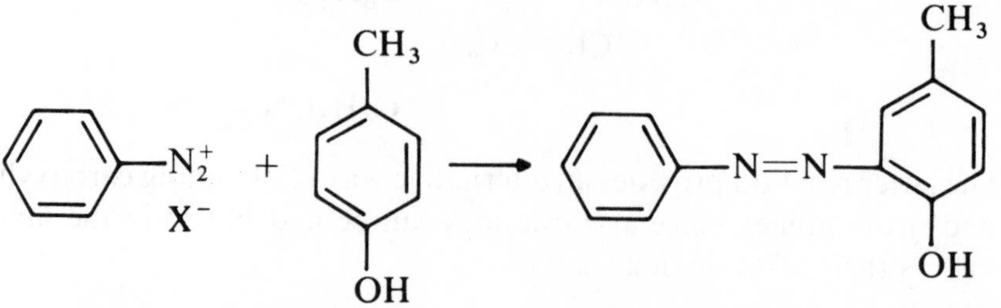

In the case of primary or secondary arylamines, diazonium salts react at the amino group instead of reacting at the *p* or *o* positions of the benzene ring:

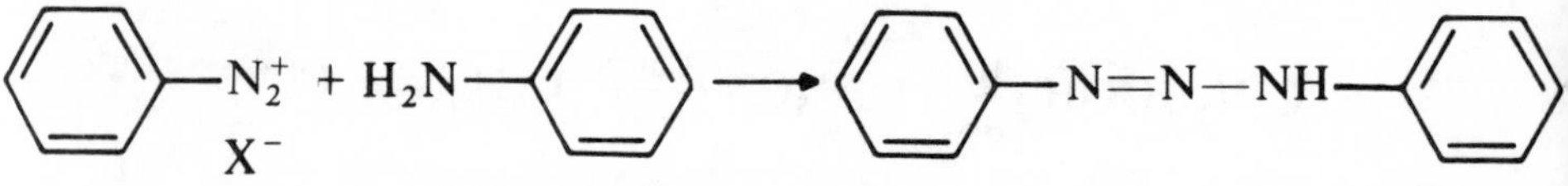

However, when these products are heated with acids they undergo a **molecular rearrangement** to give an azo compound:

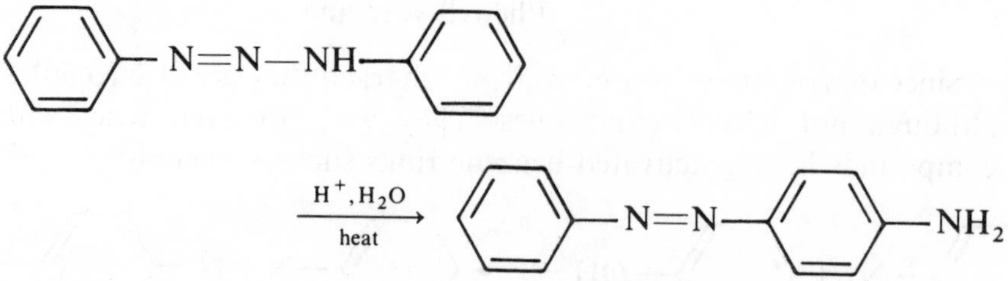

These azo compounds, made by **coupling** diazonium salts with phenols or arylamines, are of considerable commercial importance as dyestuffs, many thousands of different examples being used on fabrics and in foodstuffs and cosmetics. A wide variety of compounds, providing different shades and abilities to adhere to different materials, is obtained by using a large range of differently substituted diazonium salts, phenols and amines. Often more complex ring systems are used, providing more intense colours; e.g. the following reaction with a naphthol (see next chapter) gives a much deeper red product than is obtained from phenol:

2-Naphthol
or β-naphthol

In the azo dyeing of fabrics it is common practice to make the dyestuff actually on the material. The fabric is first impregnated with a solution of the compound that is to couple with the diazonium salt, and the latter is then added to it.

Questions

1. How may phenol be converted into (a) sodium phenoxide and (b) phenyl acetate?
2. Starting from bromobenzene, how could the following compounds be obtained: (a) benzoic acid, (b) benzyl alcohol ($C_6H_5CH_2OH$), (c) acetylbenzene and (d) 1-methyl-1-phenylethanol?
3. How could (a) p-aminotoluene and (b) benzylamine ($C_6H_5CH_2NH_2$) be obtained from toluene?
4. Starting from nitrobenzene, how could you obtain (a) cyanobenzene and (b) a dyestuff?

28 OTHER RING COMPOUNDS RELATED TO BENZENE

Naphthalene

In the discussion of diazonium salts in the previous chapter, mention was made of 2-naphthol or β-naphthol:

This compound is a phenol derived from an arene consisting of two benzene rings fused together, namely

naphthalene

Naphthalene is usually represented by the Kekulé-type formula shown. Note that it should not be represented by a double poached-egg-type formula as in

This implies the presence of twelve 'spare' electrons. Reference to the Kekulé-type formula shows that in fact it has only ten.

Naphthalene has the molecular formula $C_{10}H_8$. It is obtained from the distillation of coal-tar. It has been used for a long time domestically as mothballs. Derivatives of naphthalene have industrial importance, e.g. in the manufacture of dyestuffs.

Naphthalene and its derivatives have a chemistry that closely resembles that of benzene. For example, it undergoes electrophilic substitution reactions:

1-Bromonaphthalene
or α-bromonaphthalene

1-Nitronaphthalene
or α-nitronaphthalene

Naphthalene is more reactive in these reactions than is benzene. Electrophilic attack on benzene involves disruption of its sextet of 'spare' electrons. In the case of naphthalene a sextet is retained, and so less energy is required to bring about reaction:

which might also be written as

There are two types of CH group in naphthalene, sometimes called the α and β positions:

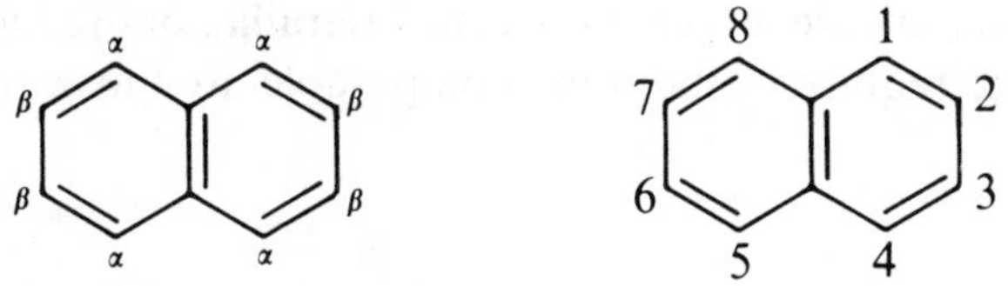

The α positions are next to the carbon atoms common to both rings, the β positions are not. Alternatively sites in the ring are indicated by numbering (for example β-naphthol $\equiv$ 2-naphthol; 1-bromonaphthalene $\equiv$ α-bromonaphthalene).

Electrophilic substitution reactions proceed preferentially at the α-or 1-position. As in the case of substituted benzenes this is because reaction there involves formation of the more stabilized intermediate:

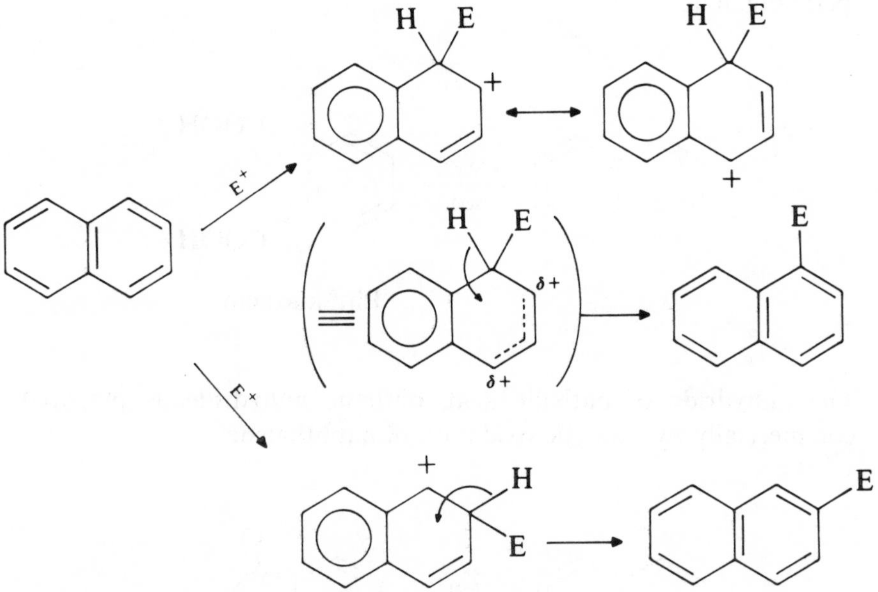

Electrophilic attack at the 1 (or α) position leads to an intermediate in which the positive charge is delocalized. This is not so in attack at the 2 (or β) position, so the latter is less favoured. (In both cases there is some small interaction between the positive charge and the sextet of 'spare' electrons in the unattacked ring.)

Attack on **one** ring of naphthalene is generally easier than on benzene, but destruction of the electron system of **both** rings is not. Thus naphthalene is more readily reduced than benzene to give a tetrahydronaphthalene (generally called **tetralin**), but reduction of the second ring requires conditions comparable to those required for benzene:

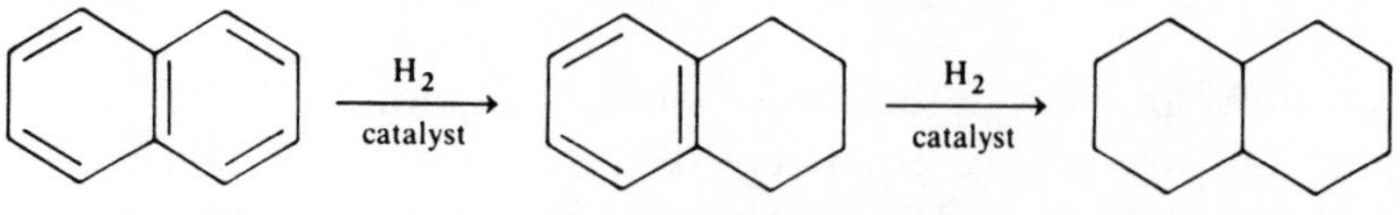

Tetrahydronaphthalene   Decahydronaphthalene
or tetralin   or decalin

Similarly, **one** of the rings can be broken by oxidation with potassium permanganate:

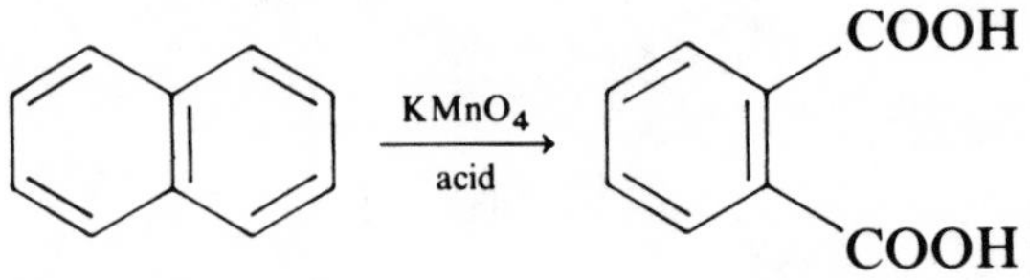

Phthalic acid

The anhydride of phthalic acid, phthalic anhydride, is prepared commercially by catalytic oxidation of naphthalene:

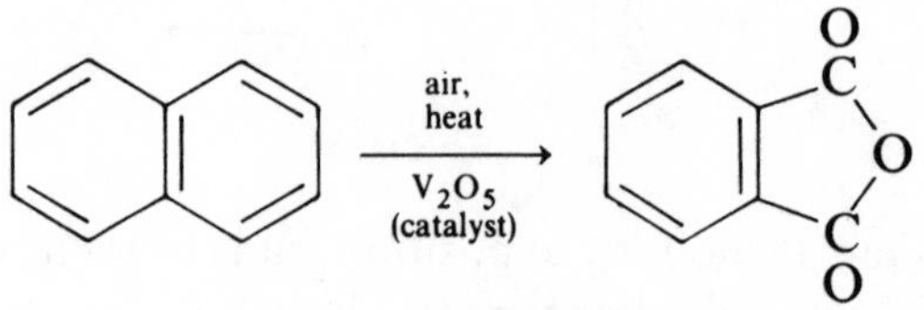

Phthalic anhydride

If three benzene rings are fused together, this can be done in two ways:

Anthracene   Phenanthrene

Both of these compounds are obtained from coal-tar; derivatives of both are of commercial importance.

Both of these isomers undergo addition reactions in the central ring:

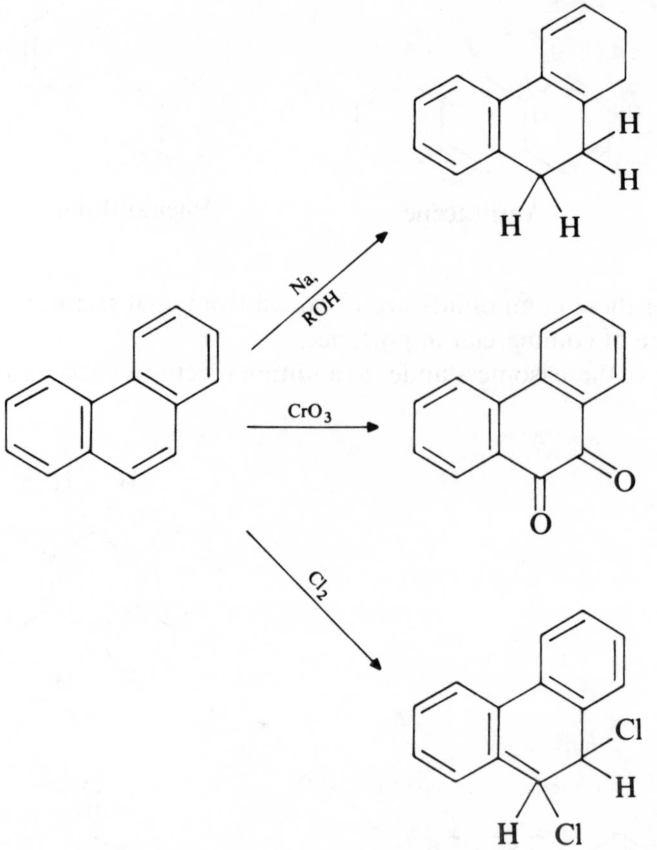

In each case two separate benzene rings are left in the product and no further reaction takes place.

There are many more complex compounds with more than three rings. Some are notorious carcinogens, i.e. cancer-causing compounds, including benzpyrene and the dibenzanthracene shown:

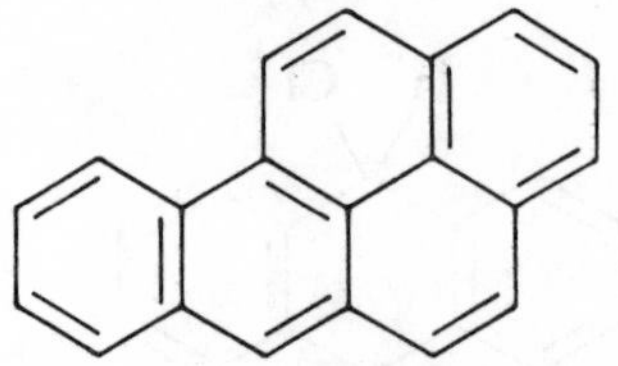

Benzpyrene

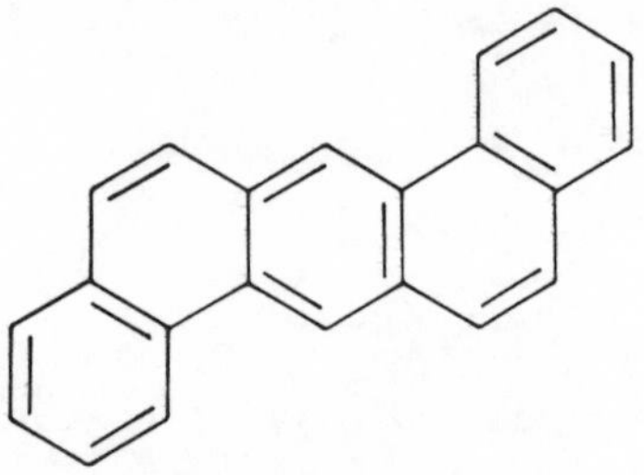

A dibenzanthracene

Many other such compounds are not carcinogens.

Some other interesting compounds having numbers of benzene rings fused together are the **helicenes**:

[5] Helicene   [6] Helicene

Helicenes having five or less rings (phenanthrene is [3]helicene) have all the rings in one plane. For helicenes with six or more rings this is not possible because the end rings overlap. The rings are therefore arranged in the shape of a screw or helix—hence their name. An end-on view of a helicene is shown (double bonds are omitted for clarity):

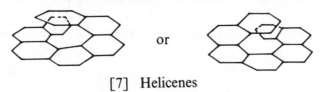

or

[7] Helicenes

The helix can be right handed or left handed, like screws. Hence this molecule is chiral and exists in mirror-image forms as shown.

Heterocyclic Compounds

The existence of heterocyclic compounds, i.e. cyclic compounds in which a ring is made up of more than one kind of atom, was pointed out earlier, in Chapter 10. They play an important role in chemistry, and make up more than three-quarters of all known organic compounds. They are of extreme importance biologically. In this connection sugars have already been mentioned (Chapter 9). One or two other examples will be mentioned briefly in the last part of this chapter.

When the rings are saturated, heterocyclic compounds closely resemble their alkyl counterparts, and mention of this has been made earlier in the chapters discussing amines and ethers.

In this chapter brief mention is made of heterocyclic analogues of benzene.

Pyridine

Pyridine has a similar structure to benzene, but one of the CH groups of benzene has been replaced by a nitrogen atom:

Pyridine

If, as in the case of benzene, each atom forming part of the ring contributes one 'spare' electron to form a delocalized sextet of electrons, then pyridine still has another two electrons over, associated with the nitrogen atom, which are not involved in bonding two atoms together. The basicity of ammonia and amines is due to their having such a 'lone pair' of electrons. Hence pyridine is also basic. This structural feature can be brought out by drawing the formula as

Pyridine forms salts with acids, and reacts, like other amines, with alkyl halides to form alkylpyridinium salts:

Pyridinium chloride

N-Methylpyridinium bromide

(The N- written before methyl in the latter name shows that the methyl group is attached to the nitrogen atom.)

It is necessary to consider whether or not pyridine resembles benzene at all in its chemical properties.

First let us consider what happens if nitration is attempted. As a base, pyridine at once reacts with the acid to form a pyridinium salt. This pyridinium cation bears a positive charge on the ring. Consequently it is only attacked with difficulty by a positively charged reactant such as an electrophile. As a result of this, reaction only takes place under very severe conditions. When pyridine is heated at 300 °C with concentrated nitric and sulphuric acids for a day, only 6 per cent of nitropyridine is formed.

The same pattern is seen with other electrophiles, which react with the lone pair of electrons on the nitrogen atom, thereby creating a positive charge on the ring, which makes it resistant to further attack by the electrophile:

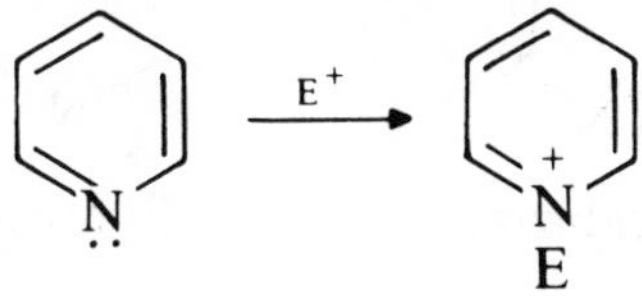

for example

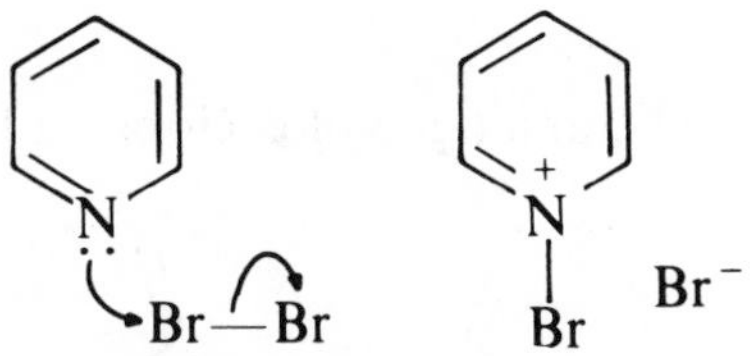

Therefore pyridine is very unreactive towards electrophiles.

The pyridine ring is very stable. When the corresponding analogue of naphthalene, **quinoline**, is oxidized, it is the carbon ring that is destroyed rather than the pyridine ring:

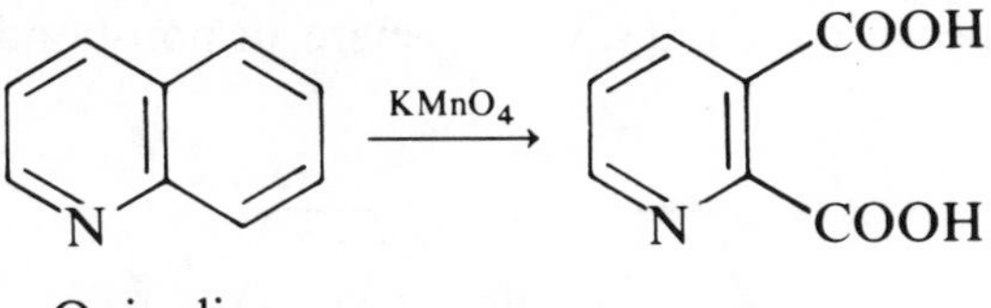

Quinoline

Indeed, pyridine can be used as a solvent for oxidizing agents such as chromium trioxide.

It is possible to replace more than one CH group of the benzene ring by nitrogen atoms, as, for example, in **pyrimidine**:

Pyrimidine

Pyrimidine is a particularly important compound because of the wide occurrence of its derivatives in biological systems, especially in nucleic acids, and in RNA (ribonucleic acid) and DNA (deoxyribonucleic acid).

There are also two isomers of pyrimidine with 1,2- and 1,4-arrangements of the nitrogen atoms:

Pyridazine   Pyrazine

Pyrrole

Pyrrole is a heterocyclic analogue of cyclopentadiene:

Cyclopentadiene   Pyrrole

In Chapter 24 it was seen that cyclopentadiene forms a particularly stable anion because it has a sextet of 'spare' or non-bonding electrons in the ring:

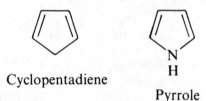

The neutral pyrrole molecule has such a sextet of non-bonding electrons in the molecule, four from the two double bonds and the lone pair of electrons normally associated with a nitrogen atom:

Hence this molecule should gain stabilization from this sextet of electrons.

Because the nitrogen atom of pyrrole shares its lone pair of electrons to provide this sextet, pyrrole is only very weakly basic.

Strong acids protonate pyrrole, as they do benzene:

In the process the sextet of electrons is lost, and in strong acid pyrrole polymerizes very readily.

Pyrrole is more electron-rich than benzene since it has six non-bonding electrons spread over only five atoms instead of over six atoms as in benzene. It undergoes electrophilic substitution very readily, but reactions involving acids are obviously ruled out since the acid would bring about polymerization. Some examples showing pyrrole's high reactivity are as follows:

Neither of these reactions needs a catalyst, as is required in the corresponding reactions of benzene.

Pyrrole and its derivatives are also highly important in nature; e.g. chlorophyll and haemoglobin, pigments of, respectively, plants and blood, are complex derivatives of pyrrole.

Furan and Thiophen

Other heteroatoms that have lone pairs of electrons can, like the NH group of pyrrole, provide two electrons in a five-membered ring to produce a compound stabilized by the presence of six 'spare' electrons. Examples are **furan** and **thiophen**:

Furan   Thiophen

Both are more reactive to electrophiles than is benzene, but less so than pyrrole. Reactivity is in the order.

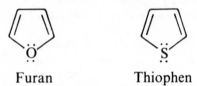

Stability to acid is in the opposite order; furan polymerizes in the presence of acids. Thiophen is very similar to benzene in its properties, although more reactive.

Thiophen is made commercially from a cyclo-addition reaction of butadiene with suphur:

$$\text{butadiene} + S \xrightarrow{600\,°C} \text{thiophen} \quad S$$

Furan is a highly important industrial chemical. It is reduced to tetrahydrofuran (see Chapter 10), which is a valuable solvent, but is also a precursor from which the dicarboxylic acid and diamine required for the preparation of nylon are made. It is readily available from waste products of the food industry such as oat hulls, rice hulls and corncobs, and is made in very large quantities from these sources. It is derived from certain sugars present in these materials.

As with six-membered rings, more than one heteroatom may be present in five-membered rings having six 'spare' electrons. Derivatives of many of these compounds, e.g. **imidazole** and **thiazole**, are also important in biological systems:

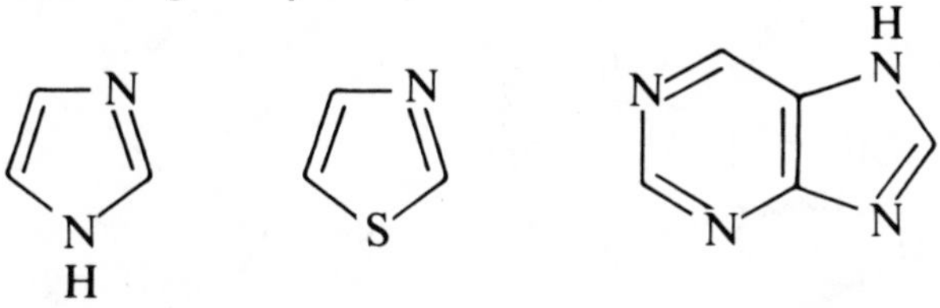

| Imidazole | Thiazole | Purine |

Derivatives of purine, in which an imidazole ring is fused to a pyrimidine ring, are also vitally important compounds in nature, forming part of nucleic acids and nucleotides, and being vital in fundamental metabolism, including heredity and evolution.

Many other more complex heterocyclic systems also play essential roles in biological processes. A good example is provided by the molecule of vitamin B12, whose structure, given on next page, provides a fitting climax to this text.

Questions

1. Write equations showing the products obtained from the following reactions, also giving the mechanisms of the reactions:
   (a) bromination of naphthalene,
   (b) reaction of pyridine with methyl iodide.
2. Give the formulae of the products obtained when (a) naphthalene, (b) anthracene and (c) phenanthrene are oxidized.
3. What reactions occur when (a) pyridine and (b) pyrrole are treated with acid? Why do they differ in their behaviour towards acids?

Vitamin B_{12}

APPENDIX
THE USE OF PHYSICAL METHODS IN THE
STUDY OF ORGANIC COMPOUNDS

For the first century and more of the study of organic chemistry, information concerning the molecular structure, i.e. the arrangements of the component atoms within the molecules, was obtained almost exclusively from studies of the chemical reactions that compounds undergo and from the types of products obtained from the reactions. Final confirmation of molecular structures most often involved synthesis of the compounds by what were hoped to be unequivocal methods.

The middle of the twentieth century saw a complete change in the methodology employed for determination of structures. While chemical methods are not completely ignored, most information is now obtained by physical methods and, in particular, from spectroscopy. The following pages are intended to give no more than a brief resumé of the most commonly used techniques. An array of books dealing with spectroscopy, at elementary as well as at more sophisticated levels, is available, and readers should turn to these texts for details of theory and practice.

Spectroscopy

It has long been known that a prism separates white light out into a range of different colours called a spectrum. These different colours have different wavelengths or frequencies, just as different radio stations do. The human eye is sensitive only to radiation (light) of a very restricted range of wavelengths, but a much wider spectrum of radiation exists which is invisible to the human eye. Thus so-called ultraviolet light has a shorter wavelength, or greater frequency, than visible light, while so-called infrared light, and radio waves, have a longer wavelength, or lesser frequency.(Frequency has a reciprocal relationship to wavelength.) Light is a form of energy and the energy is related to the wavelength or frequency by the following equations:

$$E = h\nu = \frac{hc}{\lambda}$$

(where E = energy, h = Planck's constant, ν = frequency, λ = wavelength and c = velocity of light).

Molecules absorb radiation of specific frequencies, and in so doing gain energy. This results in changes in the molecules; e.g. the absorbed energy may be converted into rotational or vibrational energy in the molecule, or alternatively the absorption of energy may result in an electron within the molecule being promoted to a level of higher energy. The particular frequencies that different molecules can absorb depend upon the atomic and electronic structures of the molecules. Thus, if the particular frequencies absorbed by an organic compound can be determined, this in turn provides detailed information about the atomic and/or electronic structure of this compound. Such information can be obtained by using appropriate spectrophotometers; these instruments present the information as a continuous plot showing the absorption at every frequency over a particular range. The form of the spectrum so obtained is dependent upon the molecular structure of the compound.

Visible and Ultraviolet Spectra (Electronic Spectra)

The visible spectrum extends from about 400 to 800 mm (1 nm = 10^{-9} m). The so-called near-ultraviolet extends from 400 to 200 nm. The limits set for 'visible' light are obvious. The limit set for near-ultraviolet light (~ 200 nm) is also determined by practical considerations. Oxygen absorbs light strongly at wavelengths below 210 nm, thus special apparatus, excluding air, is necessary to record spectra at wavelengths lower than 210 nm.

Visible and near-ultraviolet spectra are usually recorded with one instrument. They are often called electronic spectra since the major features are caused by excitation of electrons from their lowest energy or ground state to an excited state by the absorbed light. Spectra are normally recorded using solutions of the compound and information is provided both by the wavelengths at which maximum absorption (λ_{max}) takes place and by the intensity of absorption at these maxima. A typical ultraviolet spectrum is as follows:

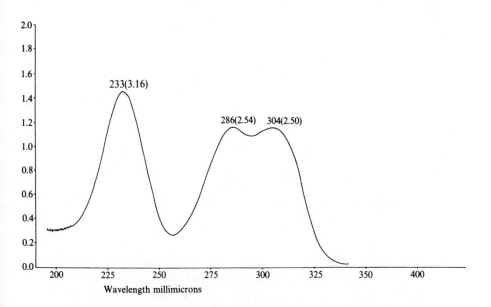

These spectra usually provide information about the overall electronic structure of the molecule, in particular about the extent of conjugation (see page 240). For example, in the case of a molecule known to contain both a carbon–carbon double bond and a carbonyl group it would show whether or not these groups were 'conjugated' with one another (see page 243). In general, however, electronic spectra do not provide evidence concerning specific structural features.

Electronic spectra provide a very valuable tool for quantitative analytical determination of compounds, by comparing the intensity of absorption of the analysed sample with that of a pure sample. This quantitative aspect has also been utilized in studying the rates at which reactions proceed, by recording either the decrease with time of intensity of an absorption maximum associated with the reactant or the increase with time of intensity of a maximum associated with the product of the reaction.

Infrared Spectra

A typical infra-red spectrum appears as follows:

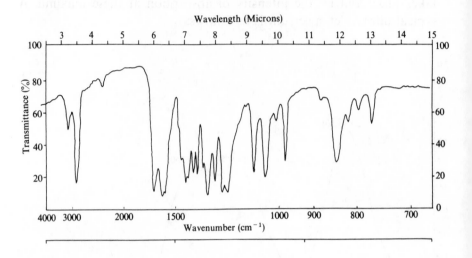

It is evident that such spectra have a much more complex appearance than do electronic spectra. They are consequently much more characteristic than are electronic spectra and are much more informative about the atomic structure of molecules than are electronic spectra.

It may be noted that, conventionally, the vertical scale for electronic spectra records **absorbance**, whereas that for infrared spectra records **transmittance**; absorption maxima are presented upward in the former but downward in the latter.

The region of the spectrum that is commonly observed is that from 2.5 to 15.0 μm (1 μm = 10^{-6} m). An alternative convention uses wavenumbers instead of wavelengths. Wavenumbers ($\bar{\nu}$) are related to wavelengths (λ) by the relationship $\bar{\nu} = 10^4/\lambda$, and are given in units of reciprocal centimetres, so that on this scale the normal range is 4000–650 cm^{-1}. Each of these conventions is in common usage and both are normally given on the spectrum chart. Infrared spectra are most commonly recorded in the case of a liquid by using a neat sample. In the case of a solid, either a mull is made by grinding the sample with liquid paraffin (which, however, itself absorbs in some regions of the spectrum and will thus hide other absorptions in those regions) or other suitable liquids, or a sample is ground with dry solid potassium bromide and the resultant mixture is compressed into a disc; the mull or disc is then put into the spectrophotometer.

Organic molecules are in a constant state of vibration; the bonds between the atoms are continually stretching and bending to and fro. Each bond in a molecule has a characteristic stretching and bending frequency and can absorb radiation of that frequency. The frequencies involved are those associated with infrared radiation. Particular groups of atoms give rise to absorption bands at about the same frequency (wavelength) whenever they are present in a molecule, e.g. aliphatic carbonyl groups ($C=O$) at about $1755-1695 \, cm^{-1}$ or $5.70-5.90 \, \mu m$. Thus an infrared spectrum may indicate the strong probability that a carbonyl group is present in a molecule—not the absolute certainty, because it is always possible that some other less likely group is providing the absorption band, but further information from other sources will usually settle the matter. However, the absence of absorption in the appropriate region is very strong evidence for the absence of a particular group of atoms.

The exact location of an absorption peak that is associated with a structural feature may provide more detailed information about the environment of that feature; e.g. an aliphatic ketone (RCOR) usually provides an absorption in the range $1725-1705 \, cm^{-1}$ ($5.80-5.87 \, \mu m$), whereas an aliphatic aldehyde (RCHO) usually provides a peak at about $1740-1720 \, cm^{-1}$ ($5.75-5.81 \, \mu m$) and an aliphatic ester (RCOOR') at $1750-1735 \, cm^{-1}$ ($5.71-5.76 \, \mu m$). Similarly a carbonyl group which forms part of a saturated six-membered ring, as in cyclohexanone, absorbs at $1725-1705 \, cm^{-1}$, whereas if it forms part of a five- or four-membered ring it absorbs at $1750-1740$ or $1780-1760 \, cm^{-1}$ respectively.

Another invaluable use of infrared spectra is in proving the identity of two samples. Proof of the identity of any organic compound rests upon the accumulation of physical evidence. An infrared spectrum is really a vast collection of physical data for it is the record of the absorption of radiation at a very large number of different wavelengths. Normally infrared spectra are much more irregular and have more conspicuous features than electronic spectra. If two samples of material provide identical infrared spectra, it is as near as possible certain that they consist of the same compound.

Nuclear Magnetic Resonance (n.m.r.) Spectra

Under appropriate conditions a sample of material can absorb radiation in the radio-frequency region, say at $\sim 100\,\mathrm{MHz}$. This absorption is dependent upon the magnetic moments associated with nuclear particles. The experiment, most commonly carried out with a solution of the sample, consists of irradiating this sample with radio-frequency radiation while it is held in a powerful magnetic field. Nuclei in different chemical environments absorb at slightly different frequencies of radiation when the magnetic field is kept constant. In practice a reference compound is often also included in the solution, most commonly tetramethylsilane (TMS), $[(CH_3)_4Si]$, and n.m.r. spectra record the positions of absorption signals in dimensionless units, δ, which record the separation of the absorption peaks of the sample being examined from the peak due to TMS. Because this separation is due to the chemical environment of the atoms under study it is described as the **chemical shift**. The term **nuclear magnetic resonance (n.m.r.) spectrum** reflects its association with the magnetic properties of atomic nuclei.

Nuclei having an even mass number and an even atomic number, for example ^{12}C and ^{16}O, are non-magnetic and therefore do not provide n.m.r. spectra. However, protons do provide n.m.r. spectra and, since most organic compounds contain hydrogen, they in turn provide ^1H-n.m.r. spectra. Although ^{12}C is non-magnetic, ^{13}C is not, and, since naturally occurring carbon contains 1.1% ^{13}C, spectra representing the carbon atoms of organic compounds are readily obtained, although an instrument of higher sensitivity is required. Other elements are also studied using this technique, but as far as organic chemistry is concerned, ^{13}C-n.m.r. and ^1H-n.m.r. spectra are obviously of paramount importance. ^{13}C-n.m.r. and ^1H-n.m.r. spectra are recorded separately.

The chemical shift associated with an atom depends essentially on its chemical environment. Thus n.m.r. spectra provide (in theory, at least; in practice there may sometimes be difficulties because of overlapping signals) measures of the numbers of different environments of ^1H and ^{13}C atoms in a molecule; in addition the extent of the chemical shift provides information about the nature of the environment. Furthermore, in the case of ^1H-n.m.r. spectra quantitative data are obtained, signifying the ratios of the numbers of protons in each of the environments; this is not normally the case with routine ^{13}C-n.m.r. spectra.

As an example, when the ^{13}C-n.m.r. and ^{1}H-n.m.r. spectra of acetone (CH$_3$COCH$_3$) are recorded there is only one signal in the ^{1}H spectrum but two separate signals in the ^{13}C spectrum. All the hydrogen atoms are equivalent and in the same environment; they belong to two methyl groups each of which is attached to a carbonyl group. Two carbon atoms form part of these methyl groups and are equivalent, but the third carbon atom belongs to the carbonyl group; it is thus in a different environment from the other two carbon atoms and there are consequently two carbon signals. In the case of pentan-3-one (CH$_3$CH$_2$COCH$_2$CH$_3$) there are two signals in the ^{1}H-n.m.r. spectrum, with relative sizes 2:3; six hydrogen atoms belong to two identical methyl groups and four more to two identical methylene (CH$_2$) groups. The ^{13}C-n.m.r. spectrum has three signals corresponding to the methyl, methylene and carbonyl groups.

Also for technical reasons the presentations of ^{13}C-n.m.r. and ^{1}H-n.m.r. spectra are different. In the case of acetone they appear as follows:

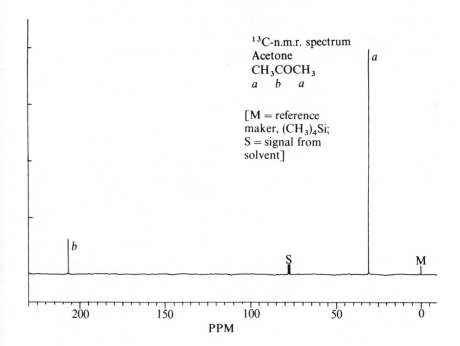

^{13}C-n.m.r. spectrum
Acetone
CH$_3$COCH$_3$
a   b   a

[M = reference maker, (CH$_3$)$_4$Si; S = signal from solvent]

316   A First Course in Organic Chemistry

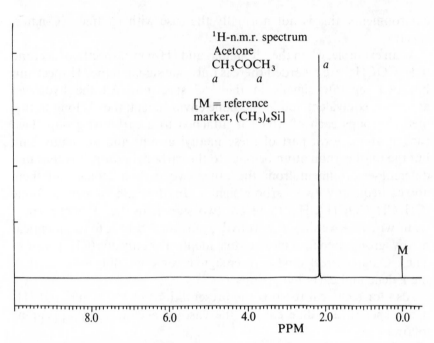

The carbon signals normally appear as sharp lines, the proton signals as peaks, which, depending on the type of environment of the proton(s) involved, may be very narrow or broad.

In the spectra of pentan-3-one a further difference is evident:

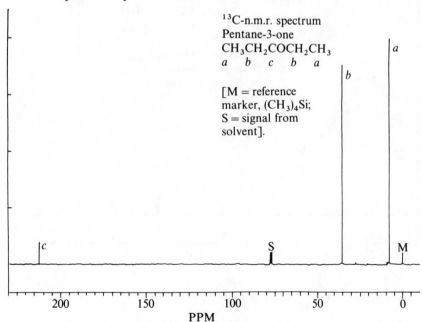

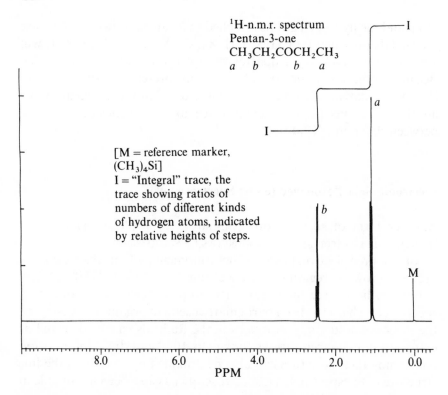

¹H-n.m.r. spectrum
Pentan-3-one
$CH_3CH_2COCH_2CH_3$
 a   b   b   a

[M = reference marker,
$(CH_3)_4Si$]
I = "Integral" trace, the
trace showing ratios of
numbers of different kinds
of hydrogen atoms, indicated
by relative heights of steps.

8.0   6.0   4.0   2.0   0.0
                    PPM

In the case of the ¹³C-n.m.r. spectrum three single lines appear, but in the ¹H-n.m.r. spectrum, although there are obviously two signals, these signals have structure, that corresponding to the CH_2 group being a **quartet**, with four distinct sub-peaks, and that corresponding to the CH_3 groups being a **triplet**, with three distinct sub-peaks. This phenomenon is known as **splitting**. It will not be discussed in detail here because it can be complicated, but it is associated with interactions, described as **coupling**, between hydrogen atoms attached to neighbouring carbon atoms. In very simple cases the peak is split into sub-peaks which number $(n + 1)$, where n is the number of hydrogen atoms attached to the next carbon atom(s). Thus in the above case the CH_3 groups provide a triplet signal because in each case the adjacent carbon atom bears two (identical) protons, whereas the CH_2 groups provide a quartet because the adjacent methyl groups in each case bear three protons. The splitting patterns of ¹H-n.m.r. spectra can provide valuable information about the structure of the compound in question.

Because of the manner in which they are acquired in the spectrometer, routine ¹³C-n.m.r. spectra commonly do not show splitting, although by varying the mode of operation, coupling between the ¹³C atom and the hydrogen atoms attached to it can be shown. In that case the signals provided by CH_3X, CH_2X_2, CHX_3 and CX_4 (X ≠ H) will appear, respectively, as a quartet, a triplet, a doublet and a singlet. Because of the small numbers of ^{13}C atoms present in natural carbon (1.1%), it is statistically unlikely to find two ^{13}C atoms adjacent to one another in a molecule, and there is hence no need to consider coupling between different ^{13}C atoms.

Electronic Spin Resonance (e.s.r.) Spectroscopy

Another form of spectroscopy, similar in principle to n.m.r. spectroscopy, is **electron spin resonance (e.s.r.) spectroscopy.**

An unpaired electron has a magnetic moment and can absorb energy in the microwave region of the spectrum (~ 9–10×10^9 Hz).

Free radicals (see page 49) have unpaired electrons and e.s.r. spectroscopy has thus been particularly useful in organic chemistry for the detection and study of free radicals. Radicals can be detected at extremely low concentrations, down to 10^{-4} mol/dm^3. Information concerning their structure can also be obtained by a study of the fine structure of the spectra. E.s.r. spectroscopy has also been invaluable in the study of short-lived radical intermediates in organic reactions.

Mass Spectrometry

Mass spectrometry is different from the spectroscopic techniques described in the previous paragraphs. In the mass spectrometer a sample of a compound is ionized, commonly by bombardment with high-energy electrons. This causes ionization of some of the molecules giving cations. Some of these ions may also be broken apart to provide smaller molecular fragments. These fragments may be cations or neutral fragments; normally only the cations are studied. The products of this bombardment are separated magnetically according to their mass (more strictly mass/charge, but the charge is most commonly one). The so-called **parent ion** is derived from a molecule that has lost one electron, so that its mass provides a measure of the molecular weight of the molecule from which it came. Since most elements consist of mixtures of isotopes, the molecular weight obtained by many other methods is an average reflecting the distribution of all the isotopic possibilities in a gross sample of the compound. However, in mass spectrometry the molecular weights obtained are exact numerical ones; the main molecular ion peak will represent, in the case of hydrocarbons, the exact mass of the compound assembled from the most prevalent isotopic forms of carbon and hydrogen, viz. ^{12}C and ^{1}H. There will also be subsidiary peaks from molecules containing other isotopes, for example ^{13}C, but, since natural carbon and hydrogen contain, respectively, 98.89 and 99.985% of ^{12}C and ^{1}H, molecules made up exclusively of these isotopes will form the preponderant part of any hydrocarbon.

With some other elements, e.g. chlorine and bromine, more than one isotope is present in appreciable quantity. In the case of chlorine there is about one atom of ^{37}Cl for every three of ^{35}Cl, while in the case of bromine there are almost equal numbers of ^{79}Br and ^{81}Br atoms. This can prove very informative in mass spectra. Thus a sample of any compound containing one atom of bromine will provide **two** parent ion peaks separated by two mass units; for example C_2H_5Br will give peaks corresponding to molecular weights of 108 and 110. (Note that a dibromo compound will give **three** peaks in an intensity ratio $1:2:1$, e.g. for $C_2H_4Br_2$ at 186, 188, 190.)

Molecular weights so obtained may be presented either as whole numbers, as cited in the previous paragraph, or as so-called **accurate masses**, to several places of decimals. Atomic weights are in general not exact integers and, therefore, nor are molecular weights. This enables molecular formulae to be determined in addition to molecular weights. To take a very simple example, carbon monoxide (CO), nitrogen (N_2) and ethylene (ethene) (C_2H_4) all have the same integer molecular weight, 28, but their accurate masses are, respectively, 27.9949 (CO), 28.0062 (N_2) and 28.0312 (C_2H_4). Such differences are readily resolvable by a mass spectrometer.

In addition to the parent ion, mass spectra provide a variety of peaks at lower masses. The parent molecules break apart under the influence of the intense electron bombardment, and these extra peaks arise from the breakdown fragments. The study of these fragments may provide valuable clues to the structure of the original molecule, and may additionally provide a 'fingerprint' for the compound, as in the case of infrared spectra.

X-ray Crystallography

When a light shines through a material with a fine regular mesh a characteristic **diffraction pattern** is observed; this is readily noticeable when one walks towards a light carrying an umbrella and observes the light through the fabric of the umbrella. In a similar way, when X-rays pass through a crystal in which the molecules are regularly ordered, the X-rays provide a characteristic diffraction pattern. Careful (and complicated) analysis of this diffraction pattern provides detailed information about the arrangement of the atoms in the crystal, from which a three-dimensional picture of the atomic structure of the molecule can be obtained. Not only does this give a precise picture of the molecular structure of the compound, it also gives precise measurements of the distance apart of the atoms forming the molecule and of the angles between the chemical bonds in the molecule.

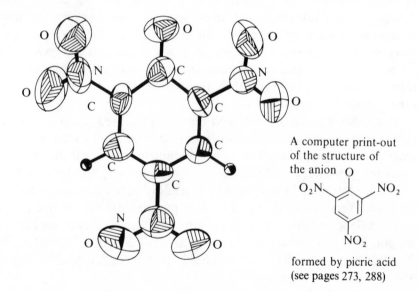

A computer print-out of the structure of the anion

formed by picric acid (see pages 273, 288)

This technique therefore provides valuable detailed information about the structure of molecules. However, the conversion of the diffraction data into molecular structures is a complex process and only very recently, with the advent of suitably sophisticated computer programs, has this become a routine tool of great value. It has limitations, however. It must be possible to obtain the compound in crystalline form at a reasonable temperature. The skill of the chemist is also required to produce good crystals of the compound suitable for examination (not to mention to make and purify the compound in the first place). Finally, it must always be borne in mind that it is possible that the structure of a molecule may change when a crystalline sample is dissolved in a solvent, in which its reactions are usually studied; fortunately this does not provide problems too often.

A related technique involves the bombardment of molecules in the gas phase by electrons. Electron diffraction patterns result and may again be interpreted in terms of molecular structure. However, **electron diffraction** is not yet as widespread a tool as is X-ray diffraction. Since the diffraction process occurs in the gas phase a limiting feature for this technique is the requirement for an adequate vapour pressure of the compound to be studied. It is an ideal technique for small molecules which provide volatile liquids or solids. Compounds having larger molecules may not be sufficiently volatile at room temperature. An increase in temperature leads to an increase in vapour pressure, and, although electron diffraction experiments have been carried out at temperature up to 1500 °C, its routine use is not yet commonplace, and there is always the risk that compounds may decompose when heated.

ANSWERS TO QUESTIONS

Chapter 2

1.

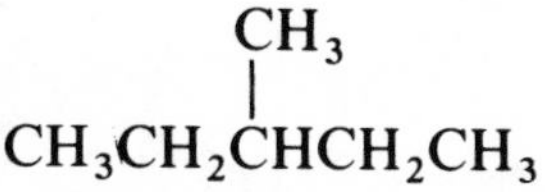

$$\underset{\displaystyle \overset{|}{CH_3}}{CH_3CH_2CHCH_2CH_3}$$

or

or

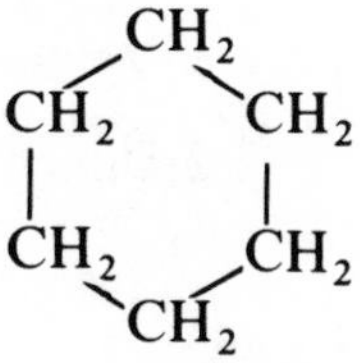

$$\underset{\displaystyle H_2C - CH_2}{CH_3 - CH \overset{\displaystyle CH_2}{\diagup} CH - CH_3}$$

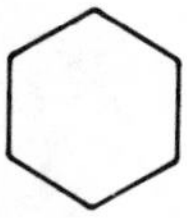

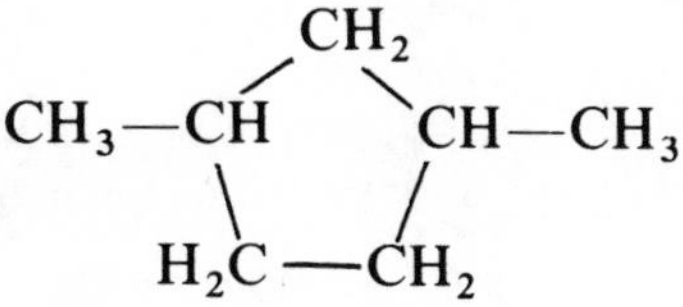

or

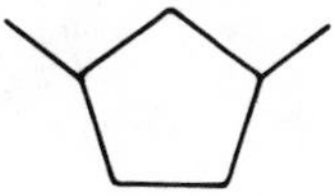

2. C_5H_{12}

$CH_3CH_2CH_2CH_2CH_3$ or    Pentane

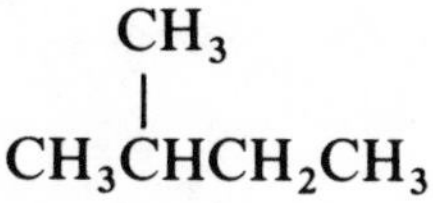

$$\underset{\displaystyle \overset{|}{CH_3}}{CH_3CHCH_2CH_3}$$ or    2-Methylbutane

$$CH_3-\underset{\underset{CH_3}{|}}{\overset{\overset{CH_3}{|}}{C}}-CH_3 \qquad \text{or} \qquad \qquad \qquad \text{2, 2-Dimethylpropane}$$

C_6H_{14}

$CH_3CH_2CH_2CH_2CH_2CH_3$ or \qquad Hexane

$$CH_3\overset{\overset{CH_3}{|}}{CH}CH_2CH_2CH_3 \qquad \text{or} \qquad \qquad \text{2-Methylpentane}$$

$$CH_3CH_2\overset{\overset{CH_3}{|}}{CH}CH_2CH_3 \qquad \text{or} \qquad \qquad \text{3-Methylpentane}$$

$$CH_3\overset{\overset{CH_3}{|}}{CH}-\overset{\overset{CH_3}{|}}{CH}CH_3 \qquad \text{or} \qquad \qquad \text{2, 3-Dimethylbutane}$$

$$CH_3\overset{\overset{CH_3}{|}}{\underset{\underset{CH_3}{|}}{CH}}CH_2CH_3 \qquad \text{or} \qquad \qquad \text{2, 2-Dimethylbutane}$$

3.

4. **Butane**

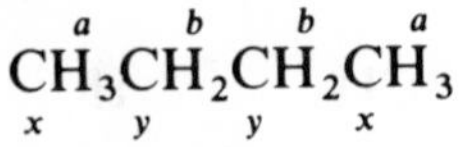

$$\underset{x}{\overset{a}{CH_3}}\underset{y}{\overset{b}{CH_2}}\underset{y}{\overset{b}{CH_2}}\underset{x}{\overset{a}{CH_3}}$$

2 kinds of H

2 kinds of C

6 of a
4 of b
x, y

2-Methylbutane

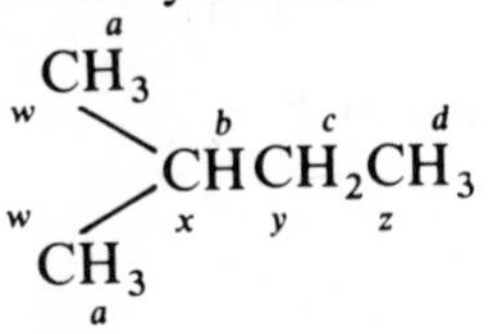

4 kinds of H

4 kinds of C

6 of a
1 of b
2 of c
3 of d
w, x, y, z

2-Methylpentane

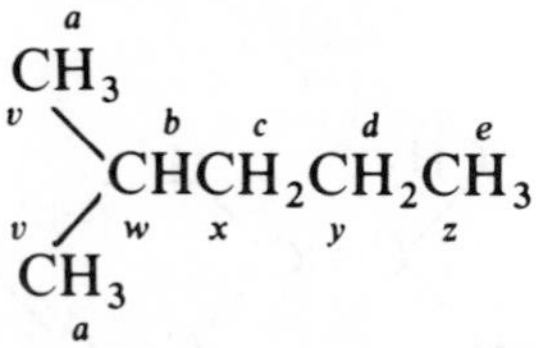

5 kinds of H

5 kinds of C

6 of a
1 of b
2 of c
2 of d
3 of e
v, w, x, y, z

2,3,4-Trimethylpentane

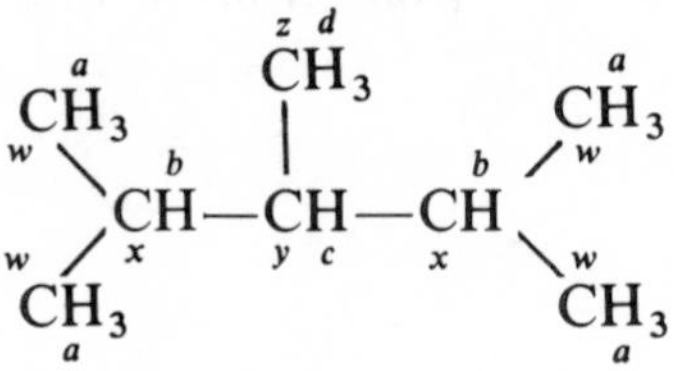

4 kinds of H

4 kinds of C

12 of a
2 of b
1 of c
3 of d
w, x, y, z

2,2-Dimethylpropane

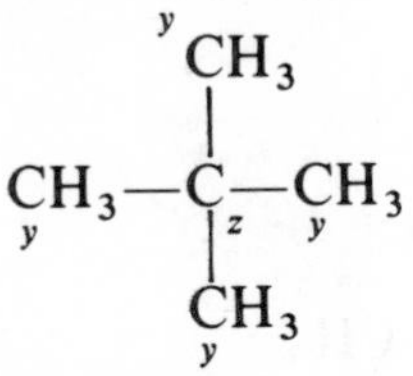

1 kind of H
2 kinds of C

y, z

Cyclopentane

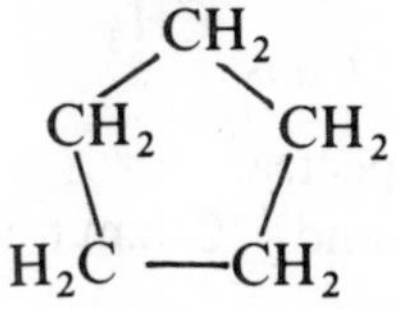

1 kind of H
1 kind of C

Chapter 3

1. (a) Two

CH₃   CH₃
  \   /
   C = C
  /   \
 H   H

cis or *Z*

CH₃   H
  \   /
   C = C
  /   \
 H   CH₃

trans or *E*

(b) one

H   H
 \   /
  C = C
 /   \
H   CH₂CH₃

(c) Three

H   H
 \   /
  C = C
 /   \
CH₃   CH₃
  \   /
   C = C
  /   \
 H   H

cis, cis or *Z, Z*

CH₃   H
  \   /
   C = C
  /   \
 H   C = C
          /   \
         H   CH₃
              H

trans, trans or *E, E*

CH₃   H
  \   /
   C = C
  /   \
 H   CH₃
      C = C
     /   \
    H   H

cis, trans or *Z, E*

(d) Two

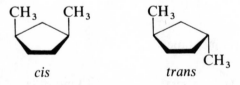

cis   *trans*

2. (a) one signal in both ¹H and ¹³C n.m.r. spectra.
   (b) and (c) both give two signals in their ¹H and ¹³C n.m.r. spectra.

3.
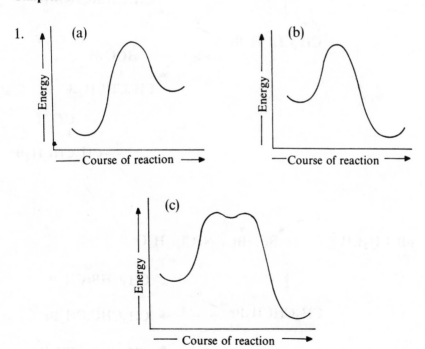

Chapter 4

1.

(a)

(b)

(c)

Chapter 5

1. (a) $CH_3CH{=\!=}CH_2 + H{-}Cl \longrightarrow CH_3\overset{+}{C}HCH_3 \overset{Cl^-}{\longrightarrow} CH_3CHClCH_3$

(b) $CH_3CH{=\!=}CH_2 + Br{-}Br + H_2O$

$$\downarrow$$

$$CH_3\overset{+}{C}HCH_2Br$$

$\overset{Br^-}{\nearrow} CH_3CHBrCH_2Br$

$\underset{H_2O}{\searrow}$
$$\overset{\overset{+}{O}H_2}{\underset{|}{CH_3CHCH_2Br}}$$

$$\longrightarrow CH_3CHOHCH_2Br$$

(c) $CH_3CH{=\!=}CH_2 + Br{-}Br + C_2H_5OH$

$$\downarrow$$

$$CH_3\overset{+}{C}HCH_2Br$$

$\overset{Br^-}{\nearrow} CH_3CHBrCH_2Br$

$\underset{C_2H_5OH}{\searrow}$
$$\overset{\overset{+}{H}OC_2H_5}{\underset{|}{CH_3CHCH_2Br}} \longrightarrow$$

$$\overset{OC_2H_5}{\underset{|}{\longrightarrow CH_3CHCH_2Br}}$$

(d) $CH_3CH{=\!=}CH_2 + Br{-}Br + NaCl + H_2O$

$$\downarrow$$

$$CH_3\overset{+}{C}HCH_2Br$$

$\overset{Br^-}{\longrightarrow} CH_3CHBrCH_2Br$

$\overset{Cl^-}{\longrightarrow} CH_3CHClCH_2Br$

$\underset{H_2O}{\searrow} CH_3CHOHCH_2Br$

2. $CH_3CH{=}CH_2 + H{-}Br \xrightarrow{\text{solution}} CH_3\overset{+}{C}HCH_3 \xrightarrow{Br^-} CH_3CHBrCH_3$

In gas phase:

$$H{-}Br \xrightarrow{hv} H\cdot \ \cdot Br$$

$$CH_3\overset{.}{C}H{=}CH_2 + \cdot Br \longrightarrow CH_3\overset{.}{C}HCH_2Br$$

$$H{-}Br$$

$$CH_3CH_2CH_2Br + \cdot Br$$

$$+ CH_3CH{=}CH_2$$

$$CH_3\overset{.}{C}HCH_2Br$$

etc.

3. $CH_2{=}CH_2 + KMnO_4 \longrightarrow HOCH_2CH_2OH$

$$CH_3CH{=}CH_2 + \text{conc. } H_2SO_4 \longrightarrow CH_3\overset{\overset{\displaystyle OSO_3H}{|}}{C}HCH_3 \xrightarrow{H_2O} CH_3CHOHCH_3$$

$$CH_3CH{=}CH_2 + Br_2 \xrightarrow{hv} CH_3CHBrCH_2Br + CH_2BrCH{=}CH_2 +$$
$$CH_2Br{-}CHBr{-}CH_2Br$$

4.

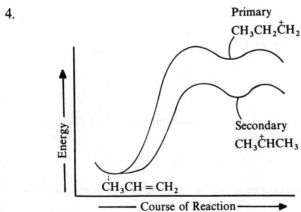

The reaction requiring the lesser activation energy is the preferred one

Chapter 6

1. (a) $C_2H_6 + Cl_2 \xrightarrow{\text{light}} C_2H_5Cl$

    Reaction takes place. Light splits the Cl—Cl bond to give Cl atoms which attack C_2H_6:

    $Cl \cdot + C_2H_6 \longrightarrow HCl + \cdot C_2H_5$
    $\cdot C_2H_5 + Cl_2 \longrightarrow C_2H_5Cl + Cl \cdot$
    etc.

    (b) No reaction takes place between C_2H_6 and KCl. KCl is an ionic reagent and alkanes do not react with such reagents.

Chapter 7

1. (a) CH_3Cl

    (b) $CH_3CHBrCH_2CH_3$

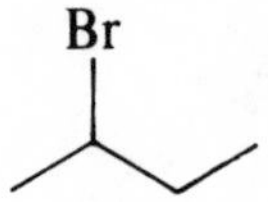

    (c)

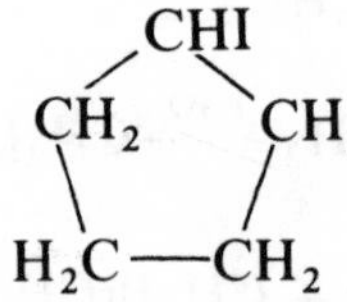

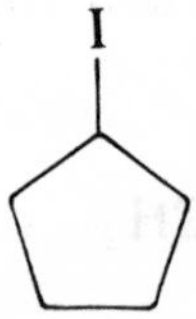

    (d)

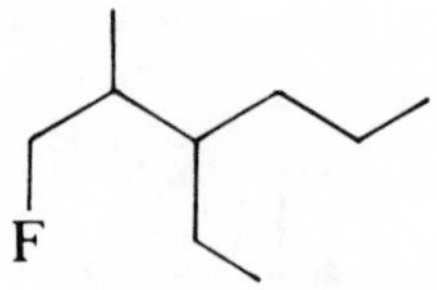

2. In homolytic cleavage of a chemical bond one electron goes to each atom at the ends of the bond:

    $$X—Y \longrightarrow X \cdot \quad \cdot Y$$

    In heterolytic cleavage of a chemical bond both electrons go to one of the atoms:

    $$X—Y \longrightarrow X \quad :Y \text{ (or } X^+ \quad Y^-)$$

Reaction (a) involves homolytic cleavage. Reaction (b) involves heterolytic cleavage.

3. (a) $RBr + HO^- \rightarrow ROH$ Alcohols
   (b) $RBr + RO^- \rightarrow ROR$ Ethers
   (c) $RCl + NH_3 \rightarrow R\overset{+}{N}H_3$ Cl^- Hydrochloride salts of primary amines

4. Use a solvent in which sodium chloride is insoluble, but sodium iodide and the alkyl chloride are soluble (e.g. acetone).

5. It would be preferable to use method (a). Method (b) leads to formation of an alkene as well as an ether:

(a) $(CH_3)_3C-O^-\ CH_3-Cl \rightarrow (CH_3)_3C-O-CH_3 + Cl^-\ Na^+$

       Na^+

(b) $(CH_3)_3CCl \rightarrow$

            $\longrightarrow H_2C=C(CH_3)_2$

and

       $\longrightarrow CH_3-\underset{\underset{CH_3}{|}}{\overset{\overset{CH_3}{|}}{C}}-OCH_3$

         CH_3O^-

          Na^+

6. $2\,C_2H_5Cl + Na_2SO_4 \rightarrow (C_2H_5)_2SO_4$

   This reaction does **not** take place. SO_4^{2-} is an anion derived from a very strong acid and is consequently a very poor nucleophile.

7. (a) $ClCH_2CH_3$ 2 signals (2 kinds of carbon atom, x, y)
       x y
   (b) $ClCH_2CH_2Cl$ 1 signal (2 identical carbon atoms)
       x x
   (c) Cl_2CHCH_3 2 signals
       x y
   (d) $ClCH_2CH_2CH_3$ 3 signals (3 different kinds of carbon atom, x, y, z)
       x y z
   (e) $CH_3CHClCH_3$ 2 signals (2 kinds of carbon atom, x, y).
       x y x

8. $$CH_2{=}CH_2 \xrightarrow[HCl]{HBr\ or} CH_3CH_2Br \xrightarrow{KCN} CH_3CH_2CN$$

$$\text{(or } CH_3CH_2Cl\text{)}$$

9. Of the isomers C_4H_9Cl only

$$\overset{a}{CH_3}$$
$$|$$
$$\overset{a}{CH_3}{-}\underset{b}{C}{-}Cl$$
$$|$$
$$\underset{a}{CH_3}$$

shows only one signal in its ^{1}H-n.m.r. spectrum (all the H atoms are equivalent) and two signals in its ^{13}C-n.m.r. spectrum (C atoms a and b).

If C_4H_8 is formed on treatment with base, an elimination reaction must occur, which would give

This would give two signals in the ^{1}H-n.m.r. spectrum (H atoms p and q) and three signals in the ^{13}C-n.m.r. spectrum (C atoms x, y and z).

Mechanisms could be:

1.

or (2)

Chapter 8

1. (a) $CH_3CH_2CH_2NH_2$
   (b) CH_3CHCH_3
   $\qquad$ |
   $\qquad$ NH_2
   (c) $CH_3CH_2NHCH_3$
   (d) $(CH_3)_3N$

2. (a) $(C_2H_5)_3N$
   (b) In the cold $(C_2H_5)_4N^+$ $\ ^-OH$;
   when heated $(C_2H_5)_3N + CH_2{=}CH_2$
   (c) $(CH_3)_2NC_2H_5$ ethyldimethylamine
   (d) $(CH_3)_3N + CH_2{=}CH_2$

Chapter 9

1. $C{:}H{:}O = 3{:}8{:}1$
   Possible structural formulae:

		$CH_3OCH_2CH_3$	$CH_3CH_2CH_2OH$	$CH_3CHOHCH_3$
Numbers of	^1H	3	4	3
n.m.r. signals	^{13}C	3	3	2

2. Molecular weight of ethyl bromide $= 109$
   Molecular weight of ethanol $= 46$
   Therefore 10.9 g of ethyl bromide should give 4.6 g of ethanol and percentage yield obtained $= 50\%$. The likely by-product is ethylene, formed in an elimination reaction.
3. (a) Alkanes do not react with bases.
   (b) $[NO_3]^-$ is a poor nucleophile, since it is an anion derived from a strong acid.
   (c) Sodium chloride is insoluble in acetone.
   (d) HO^- would preferentially remove H^+ from the N atom, that is
   $(C_2H_5)_3\overset{+}{N}H$ $\ ^-OH \longrightarrow (C_2H_5)_3N + H_2O$
   (e) $C_2H_5O^-$ is a stronger base than HO^-.

4. (a) Excess ROH $\xrightarrow{H_2SO_4}$ ROSO$_3$H $\xrightarrow{ROH}$ ROR

(b)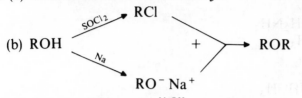

(c) CH$_3$CH$_2$CHBrCH$_3$ $\xrightarrow[H_2O]{NaOH}$ CH$_3$CH$_2$CHOHCH$_3$

$\xrightarrow{CrO_3}$

CH$_3$CH$_2$COCH$_3$

(d) CH$_3$CH$_2$OH $\xrightarrow{SOCl_2}$ CH$_3$CH$_2$Cl $\xrightarrow{KCN}$ CH$_3$CH$_2$CN

(e) CH$_3$CHOHCH$_3$ $\xrightarrow{SOCl_2}$ CH$_3$CHClCH$_3$ $\xrightarrow{NH_3}$ CH$_3$CHCH$_3$

$+$ NH$_3$ Cl$^-$

$\searrow$ NH$_3$

CH$_3$CHCH$_3$
|
NH$_2$

5. (a) CH$_3$CH$_2$CH$_2$OH $\xrightarrow[\substack{H_2SO_4, \\ heat}]{\substack{excess \\ conc.}}$ CH$_3$CH=CH$_2$

$\downarrow$ HCl

CH$_3$CHClCH$_3$

(b)
$$\underset{\underset{OH}{|}}{\overset{\overset{CH_3}{|}}{CH_3CHCH_3}} \xrightarrow[\substack{H_2SO_4 \\ heat}]{\substack{excess \\ conc.}} \overset{\overset{CH_3}{|}}{CH_3C{=}CH_2}$$

$\downarrow$ Br$_2$

$$\underset{\underset{Br}{|}}{\overset{\overset{CH_3}{|}}{CH_3C{-}CH_2Br}}$$

(c) $CH_3CHICH_3 \xrightarrow{NaNH_2} CH_3CH{=}CH_2 \xrightarrow[hv]{HBr} CH_3CH_2CH_2Br$

(d)

Chapter 10

1.

	$C_2H_5OC_2H_5$	$CH_3CH_2CH_2CH_2OH$
n.m.r.	$2\ ^{13}C$, $2\ ^1H$ signals	$4\ ^{13}C$, $5\ ^1H$ signals
i.r.	No HO peak	HO peak
+ sodium	No reaction	$\rightarrow H_2\uparrow$
+ PCl_5	No reaction	$\rightarrow HCl\uparrow$

2. Formulae: $CH_3CH_2CH_2OH$ $CH_3CHOHCH_3$ $CH_3OCH_2CH_3$
   ^1H-n.m.r.   4 signals   3 signals   3 signals
   (3:2:2:1)   (6:1:1)   (3:2:3)

Simplest distinguishing tests for one isomer are:

$CH_3OCH_2CH_3 \xrightarrow[PCl_5]{Na\ or}$ no reaction

$\left.\begin{array}{c} CH_3CH_2CH_2OH \\ or \\ CH_3CHOHCH_3 \end{array}\right\}$ $\nearrow^{Na} \quad H_2\uparrow$
$\searrow_{PCl_5} \quad HCl\uparrow$

3. No such reaction takes place. Ethers are not sufficiently polar, and $C_2H_5O^-$ is a very poor leaving group, so that nucleophilic attack by water does not take place.

4. From its molecular formula, and since it does not react with sodium, (X) must be an ether. Unsymmetric ethers react with hydrogen iodide to give two different alkyl iodides:

(X)	(Y)	(Z)
$CH_3OC(CH_3)_3 \xrightarrow{HI}$	CH_3I $+$	$IC(CH_3)_3$
3 signals in	1 signal	1 signal in
^{13}C-n.m.r.	in both	^1H-n.m.r.
2 signals (3:1)	^1H- and	2 signals in
in ^1H-n.m.r.	^{13}C-n.m.r.	^{13}C-n.m.r.

Preparation of (X):

$$(CH_3)_3COH \xrightarrow{Na} (CH_3)_3CO^- Na^+ \xrightarrow{CH_3Cl} (CH_3)_3COCH_3$$

5.

Chapter 11

1. $RCH_2CH_2X \rightarrow RCH=CH_2$

(c) HO^- attacks $R-\overset{H}{\underset{H}{C}}-\overset{H}{\underset{H}{C}}-\overset{+}{N}R'_3 \rightarrow \overset{R}{\underset{H}{>}}C=CH_2 + H_2O + NR'_3$

Chapter 12

1. $RI + Mg \xrightarrow[\text{dry ether}]{N_2,} RMgI$

$\overset{R'}{\underset{R'}{>}}C=O + RMgI \xrightarrow[\text{dry ether}]{N_2,} \overset{R'}{\underset{R'}{>}}\underset{R}{\overset{OMgI}{C}} \xrightarrow[H_2O]{H^+,} \overset{R'}{\underset{R'}{>}}\underset{R}{\overset{OH}{C}}$

2. $\overset{CH_2}{\underset{CH_2}{\|}} \xrightarrow{H_2O_2} \overset{CH_2}{\underset{CH_2}{>}}O \xrightarrow[\text{(ii) } H^+, H_2O]{\text{(i) } CH_3Li \text{ or } CH_3MgBr} CH_3CH_2CH_2OH$

3.

		Number of signals in n.m.r. spectra
	^1H	^3C
(a) $CH_3CHO \xrightarrow[\text{(ii) } H^+, H_2O]{\text{(i) } C_2H_5MgBr} CH_3-\overset{CH_2CH_3}{\underset{H}{C}}-OH$	5	4
(b) $CH_3CH_2CH_2Br \xrightarrow[\text{(ii) } H_2O]{\text{(i) } Mg/(C_2H_5)_2O/N_2} CH_3CH_2CH_3$	2	2
(c) $CH_3CH_2CH_2Br \xrightarrow[\text{(iii) } H^+, H_2O]{\overset{\text{(i)} \quad Li}{\text{(ii)} \quad CO_2}} CH_3CH_2CH_2COOH$	4	4

Chapter 14

1. $CH_3CH_2CH_2CHO$ 4 ^1H-n.m.r. signals (relative intensities 3:2:2:1)

   $CH_3CH_2COCH_3$ 3 ^1H-n.m.r. signals (3:2:3)

$$CH_3$$
$$\diagdown$$
$$CHCHO \quad 3\ ^1H\text{-n.m.r. signals } (6:1:1)$$
$$\diagup$$
$$CH_3$$

$$CH_3CH_2CH_2CH_2Cl \xrightarrow[H_2O]{NaOH,} CH_3CH_2CH_2CH_2OH$$

$$\xrightarrow{CrO_3} CH_3CH_2CH_2CHO$$

2. $CH_3COCH_3 + NH_2OH \longrightarrow CH_3CCH_3$
$$\qquad\qquad\qquad\qquad\qquad \overset{\displaystyle \|}{\underset{\displaystyle NOH}{}}$$

3. The reaction of ketones with HCN proceeds by attack of CN^- on the ketone. In acidic solution the ionization of HCN, a weak acid, is suppressed and there are insufficient cyanide ions present.

4. (a) $CH_3COCH_3 \xrightarrow{NaBH_4} CH_3CHOHCH_3$

   (b) $CH_3COCH_3 \xrightarrow{H_2NNHCONH_2} (CH_3)_2C{=}NNHCONH_2$

   (c) $CH_3COCH_3 \xrightarrow[\text{ether}]{CH_3MgBr,} CH_3\overset{\overset{\displaystyle CH_3}{\displaystyle |}}{\underset{\underset{\displaystyle OMgBr}{\displaystyle |}}{C}}CH_3 \xrightarrow[H_2O]{H^+} CH_3\overset{\overset{\displaystyle CH_3}{\displaystyle |}}{\underset{\underset{\displaystyle OH}{\displaystyle |}}{C}}CH_3$

5. (a) $CH_3CH_2OH \xrightarrow{CrO_3} CH_3CHO \xrightarrow{NaOH} CH_3CH{=}CHCHO$

   (b) $CH_3CH_2OH \xrightarrow{CrO_3} CH_3CHO \xrightarrow[\text{(ii) } H^+, H_2O]{\text{(i) } CH_3MgBr} CH_3CHOHCH_3$

(c) $BrCH_2CH_2CH_2CHO$

$\xrightarrow[H^+]{C_2H_5OH,}$ $BrCH_2CH_2CH_2CH(OC_2H_5)_2$

$\xrightarrow[H_2O]{NaOH,}$ $HOCH_2CH_2CH_2CH(OC_2H_5)_2$

$\xrightarrow[H_2O]{H^+}$ $HOCH_2CH_2CH_2CHO$

6.

(a) $CH_3\overset{\overset{\displaystyle Cl}{|}}{C}HCH_2CH_3 \longrightarrow CH_3\overset{\overset{\displaystyle }{|}}{C}HCH_2CH_3$ Nucleophilic attack

$\quad\quad\quad CN^-$ $\quad\quad\quad\quad\quad CN$

(b) $CH_3\overset{\overset{\displaystyle O}{||}}{C}CH_2CH_3 \longrightarrow CH_3\overset{\overset{\displaystyle O^-}{|}}{C}CH_2CH_3 \xrightarrow{H_2O} CH_3\overset{\overset{\displaystyle OH}{|}}{C}CH_2CH_3$

$\quad\quad\quad\quad CN^-$ $\quad\quad\quad\quad\quad CN$ $\quad\quad\quad\quad\quad CN$

Nucleophilic attack

(c) Does not react with NaCN or HBr

(d) $CH_3\overset{\overset{\displaystyle OH}{|}}{C}HCH_2CH_3 \longrightarrow CH_3\overset{\overset{\displaystyle \overset{+}{O}H_2}{|}}{C}HCH_2CH_3 \longrightarrow CH_3CHBrCH_2CH_3$

$Br\!-\!H$

$\quad\quad\quad\quad\quad\quad\quad\quad\quad\quad\quad\quad Br^-$

Electrophilic attack

(e) $(CH_3)_2CHCH\!\!=\!\!CH_2 \longrightarrow (CH_3)_2CH\overset{+}{C}H\!-\!CH_3 \longrightarrow (CH_3)_2CHCHBrCH_3$

$H\!-\!Br$

$\quad\quad\quad\quad\quad\quad\quad\quad\quad\quad\quad\quad Br^-$

Electrophilic attack

(f) $CH_3CHNH_2CH_2CH_3 \longrightarrow CH_3\overset{\overset{\displaystyle +NH_3 Br^-}{|}}{C}HCH_2CH_3$

$H\!-\!Br$

Electrophilic attack

7.

Chapter 15

1. (a) $CH_3CH_2CH_2CH_2CH_2COOH$

   (b) $CH_3CH_2\overset{\overset{\displaystyle CH_2CH_3}{|}}{C}HCH_2\overset{\overset{}{}}{C}HCH_2COOH$
   $\underset{\underset{\displaystyle CH_3}{|}}{}$

   (c) $CH_3CHBrCH_2COOH$

   (d) $CH_3CHOHCOOH$

   (e) $CH_3(CH_2)_6CH{=}CH(CH_2)_6COOH$

2. (a) $CH_3Br \xrightarrow[N_2,(C_2H_5)_2O]{Mg,} CH_3MgBr \xrightarrow[(ii)\ H^+,H_2O]{(i)\ CO_2} CH_3COOH$

   (b) $CH_3Br \xrightarrow{KCN} CH_3CN \xrightarrow{H^+,H_2O} CH_3COOH$

   $HO^-,H_2O \searrow \qquad \nearrow H^+,H_2O$

   $CH_3CO_2^-$

3. (a) $RCOOH \xrightarrow[(ii)\ H_2O,H^+]{(i)\ LiAlH_4} RCH_2OH$

   (b) $RCOOH \xrightarrow[or\ PCl_5]{SOCl_2} RCOCl$

   (c) $RCOOH \xrightarrow{N(C_2H_5)_3} RCO_2^-\ \overset{+}{N}H(C_2H_5)_3$

   (d) $RCOOH \xrightarrow[\substack{mineral\\acid,\ e.g.\\H_2SO_4}]{CH_3OH,} RCOOCH_3$

Mechanism of (d):

$$RC{\overset{\displaystyle O}{\backslash}}_{OH} \underset{}{\overset{H^+}{\rightleftharpoons}} RC{\overset{\displaystyle OH}{\lessgtr}}^+_{OH} \rightleftharpoons R\overset{\overset{\displaystyle OH}{|}}{C}{-}OH \rightleftharpoons R\overset{\overset{\displaystyle OH}{|}}{C}{-}\overset{+}{O}H_2$$

$$CH_3\overset{..}{O}H \qquad \underset{+}{CH_3\overset{..}{O}H} \qquad \underset{\displaystyle OCH_3}{|}$$

$$RC{\overset{\displaystyle O}{\backslash}}_{OCH_3} \qquad +H^+ \rightleftharpoons R\overset{\overset{\displaystyle O-H}{|}}{C}+\quad +H_2O$$
$$\rightleftharpoons \qquad \qquad \qquad \underset{\underset{\displaystyle OCH_3}{|}}{}$$

Chapter 16

1.(a)

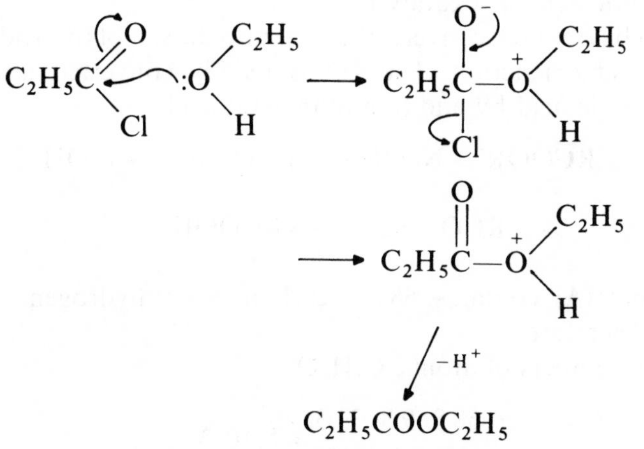

(b)

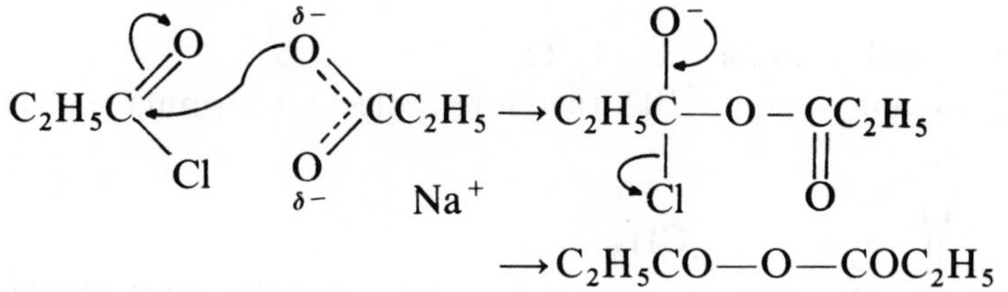

(c) $C_2H_5COCl \xrightarrow[\text{pyridine}]{\text{NaBH}_4 \text{ in}} C_2H_5CHO$

The simplified mechanism is

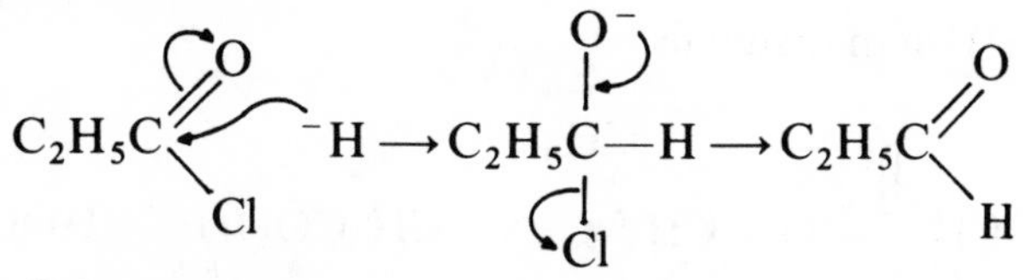

2. (a) $(CH_3CO)_2O \xrightarrow{H_2O} CH_3COOH$

(b) $(CH_3CO)_2O \xrightarrow[H_2O]{KOH} CH_3CO_2^- K^+$

(c) $(CH_3CO)_2O \xrightarrow{CH_3NH_2} CH_3CONHCH_3$

Chapter 17

1. $K = \dfrac{[CH_3COOC_2H_5][H_2O]}{[CH_3COOH][C_2H_5OH]}$

2. (a) Use a large excess of alcohol, and a mineral acid such as sulphuric acid as a catalyst.

   (b) Use a base, which converts the ester into the salt of the acid in an irreversible reaction. The salt is readily converted into the carboxylic acid by addition of mineral acid:

$$RCOOR' + NaOH \longrightarrow RCO_2^- \, Na^+ + R'OH$$

$$RCO_2^- \, Na^+ \xrightarrow[H_2O]{H^+} RCOOH$$

3. Compound (A) contains 58.8% carbon, 9.8% hydrogen, 31.3% oxygen. Therefore

   Ratios of numbers of atoms, C:H:O =

$$\frac{58.8}{12} : \frac{9.8}{1} : \frac{31.3}{16} = 5:10:2$$

Empirical formula = $C_5H_{10}O_2$

Molecular weight = 102, therefore molecular formula = $C_5H_{10}O_2$.

^{1}H-n.m.r. signals:

^{13}C-n.m.r. signals:

Mechanism of hydrolysis:

$$CH_3\overset{\overset{\displaystyle O}{\|}}{C}-OCH(CH_3)_2 \xrightarrow{H^+} CH_3\overset{\overset{\displaystyle OH}{|}}{\underset{+}{C}}-OCH(CH_3)_2$$

$$\overset{\overset{\displaystyle \ddot{O}-H}{|}}{\underset{\displaystyle H}{}}$$

$$\rightarrow CH_3\overset{\overset{\displaystyle OH}{|}}{C}-OCH(CH_3)_2$$

$$+\overset{+}{O}H_2$$

$$CH_3COOH \leftarrow CH_3\overset{\overset{\displaystyle O-H}{|}}{\underset{\displaystyle OH}{\overset{+}{C}}} + \; + HOCH(CH_3)_2 \leftarrow CH_3\overset{\overset{\displaystyle OH}{|}}{\underset{\displaystyle OH}{C}}-\overset{+}{O}\overset{\displaystyle H}{\underset{\displaystyle CH(CH_3)_2}{}}$$

(Note that all these steps are in fact equilibria.)

Chapter 19

1. $C_2H_5COCl \xrightarrow{NH_3} C_2H_5CONH_2 \xrightarrow{LiAlH_4} C_2H_5CH_2NH_2$

2. (a) Polyesters

$$+ HO \sim\!\!\sim OH + HOOC \sim\!\!\sim COOH + HO \sim\!\!\sim OH + HOOC \sim\!\!\sim COOH$$

$$\downarrow$$

$$\sim\!\!\sim COO \sim\!\!\sim OOC \sim\!\!\sim\!\!\sim COO \sim\!\!\sim OOC \sim\!\!\sim COO \sim\!\!\sim$$

A polyester

(b) Polycarbonates

$$+ HO \sim\!\!\sim OH + HO-\overset{\overset{\displaystyle }{}}{\underset{\overset{\displaystyle \|}{O}}{C}}-OH + HO \sim\!\!\sim OH + HO-\overset{}{\underset{\overset{\displaystyle \|}{O}}{C}}-OH +$$

$$\downarrow$$

$$\sim\!\!\sim O-CO-O \sim\!\!\sim O-CO-O \sim\!\!\sim O-CO-O \sim\!\!\sim\!\!\sim O-CO-O \sim\!\!\sim$$

A polycarbonate

(c) Polyamides

$$+ H_2N \sim\sim\sim NH_2 + HOOC \sim\sim\sim COOH + H_2N \sim\sim\sim NH_2 +$$

$$\downarrow$$

$$H_3\overset{+}{N} \sim\sim\sim \overset{+}{N}H_3 + \bar{O}_2C \sim\sim\sim CO_2^- + H_3\overset{+}{N} \sim\sim\sim \overset{+}{N}H_3 +$$

$$\downarrow$$

$$\sim\sim\sim CONH \sim\sim\sim NHCO \sim\sim\sim CONH \sim\sim\sim NHCO \sim\sim\sim$$

3.   A polyamide

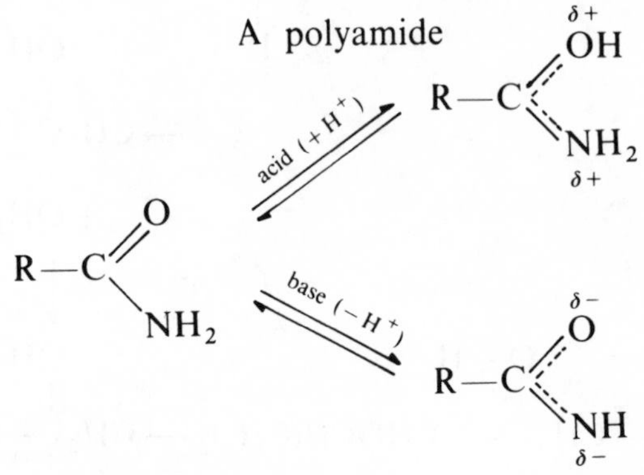

Chapter 20

1. (a) 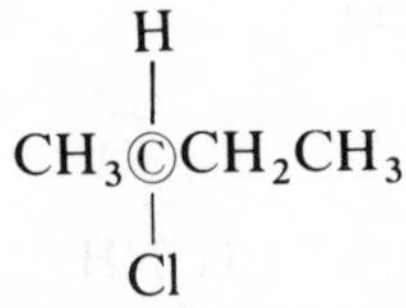 $CH_3\overset{\overset{\text{H}}{|}}{\underset{\underset{\text{Cl}}{|}}{C}}CH_2CH_3$

   (b) 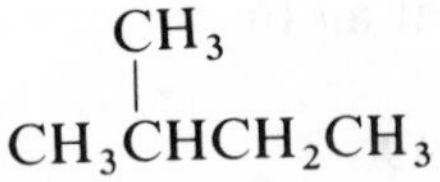 $CH_3\overset{\overset{\text{CH}_3}{|}}{C}HCH_2CH_3$

   No chiral centre

   (c) 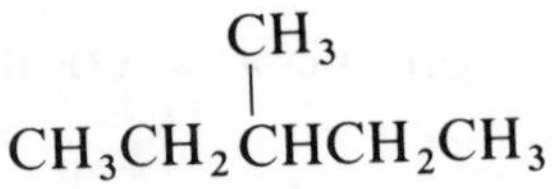 $CH_3CH_2\overset{\overset{\text{CH}_3}{|}}{C}HCH_2CH_3$

   No chiral centre

   (d) 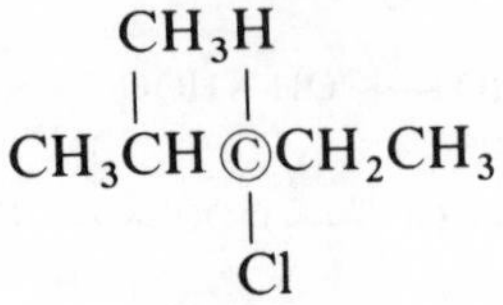 $CH_3\overset{\overset{\text{CH}_3}{|}}{C}H\overset{\overset{\text{H}}{|}}{\underset{\underset{\text{Cl}}{|}}{C}}CH_2CH_3$

   (e) 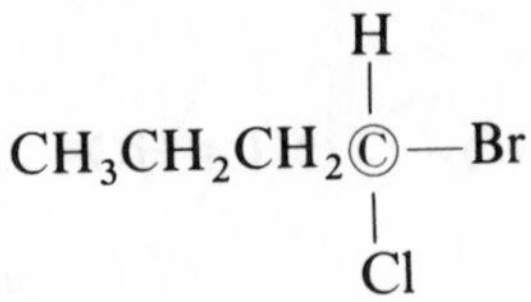 $CH_3CH_2CH_2\overset{\overset{\text{H}}{|}}{\underset{\underset{\text{Cl}}{|}}{C}}-Br$

   (f) $CH_3-\overset{\overset{\text{Br}}{|}}{\underset{\underset{\text{H}}{|}}{C}}-\overset{\overset{\text{Br}}{|}}{\underset{\underset{\text{H}}{|}}{C}}-CH_2CH_3$

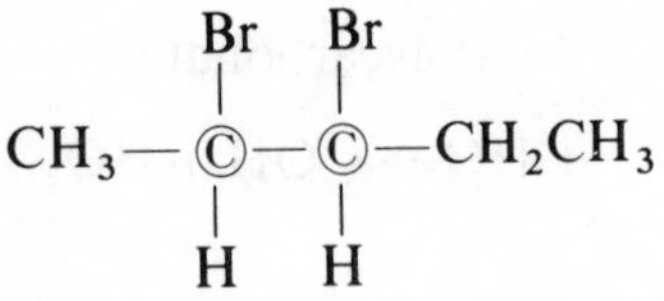

2.

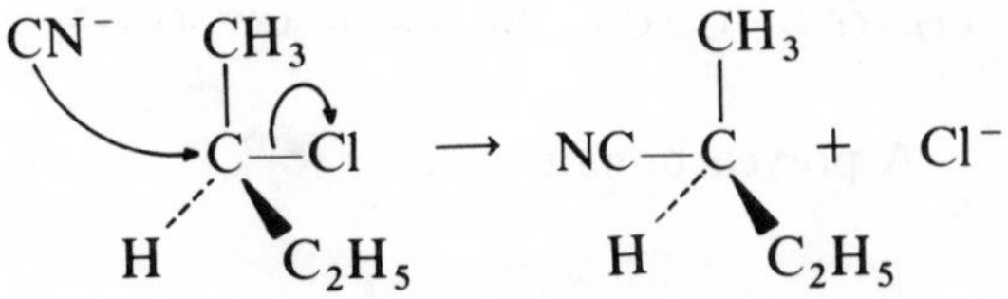

(This can also be represented in other ways; the important point is to show that a right-handed or left-handed isomer gives a product with the opposite handedness, resulting from the 'rearward' attack of CN^- on the C—Cl bond.)

3. (a) RCOOCH$_2$
      |
   RCOOCH
      |
   RCOOCH$_2$

(b) ∼ CONH—CHR—CONH—CHR′—CONH—CHR″—CONH ∼∼

4.

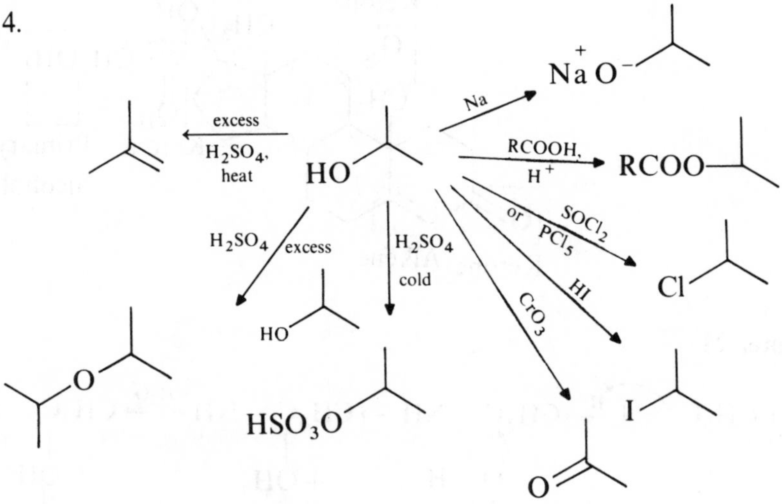

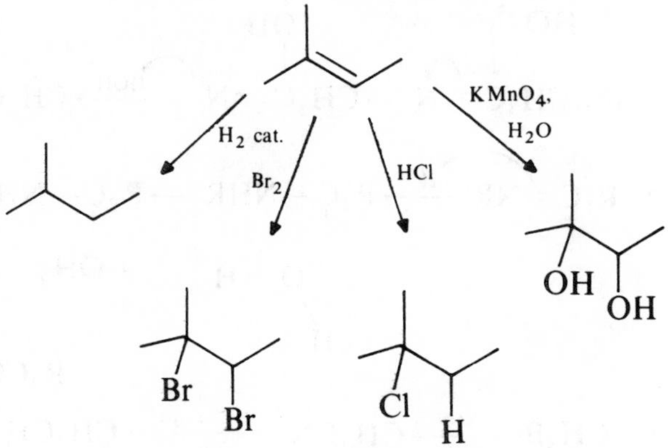

5.

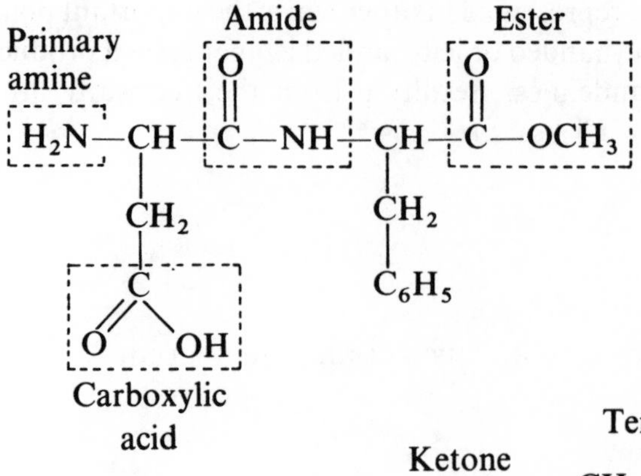

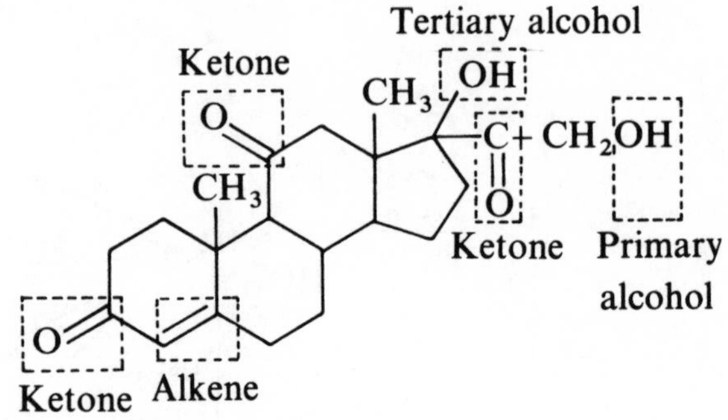

Primary amine / Amide / Ester

$$H_2N-CH-C-NH-CH-C-OCH_3$$

with CH_2 below first CH leading to C (O, OH) — Carboxylic acid

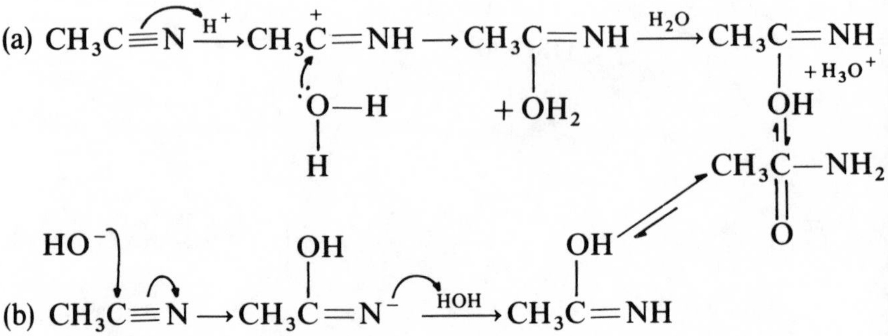

and CH_2, C_6H_5 below second CH

Ketone / Tertiary alcohol / Ketone / Primary alcohol / Alkene

Chapter 21

1. (a) $CH_3C{\equiv}N \xrightarrow{H^+} CH_3\overset{+}{C}{=}NH \rightarrow CH_3C{=}NH \xrightarrow{H_2O} CH_3C{=}NH$

$$\overset{\cdot\cdot}{O}-H \qquad +\overset{+}{O}H_2 \qquad \overset{OH}{\underset{\downarrow\uparrow}{|}} \quad +H_3O^+$$

$$H$$

$$CH_3C-NH_2$$
$$\overset{||}{O}$$

(b) $CH_3C{\equiv}N \rightarrow CH_3\overset{OH}{\underset{|}{C}}{=}N^- \xrightarrow{HOH} CH_3\overset{OH}{\underset{|}{C}}{=}NH$

with HO^- attacking

2. $R_2C{=}NR' \xrightarrow{H^+} R_2\overset{+}{C}-NHR' \rightarrow R_2C-NHR' \rightarrow R_2C-\overset{+}{N}H_2R'$

$$\overset{\cdot\cdot}{O}-H \qquad +\overset{+}{O}H_2 \qquad H-\overset{\frown}{O}$$

$$H$$

$$R_2CO + \overset{+}{N}H_3R$$

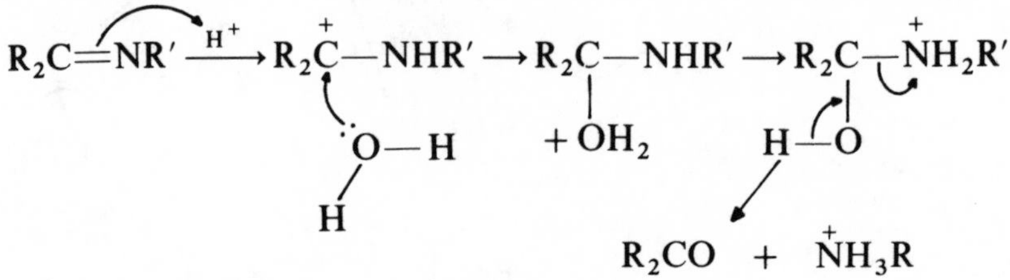

3. $CH_3Br \xrightarrow{KCN} CH_3CN \xrightarrow[\text{or LiAlH}_4]{H_2,\ \text{catalyst}} CH_3CH_2NH_2$

Chapter 22

1. $\overset{a}{C}H-\overset{b}{C}H_2$ $\overset{a}{C}H_2=\overset{b}{C}H-\overset{b}{C}H=\overset{a}{C}H_2$ $\overset{a}{C}H_3\overset{b}{C}\equiv\overset{b}{C}\overset{a}{C}H_3$
 $\underset{a}{\overset{}{C}H}-\underset{a}{\overset{}{C}H_2}$

2. (a) $HC\equiv CH \xrightarrow{Br_2} CHBr=CHBr \xrightarrow{Br_2} CHBr_2-CHBr_2$

(b) $HC\equiv CH \xrightarrow{HCl} CH_2=CHCl \xrightarrow{HCl} CH_3-CHCl_2$

(c) $HC\equiv CH \xrightarrow[HgSO_4]{H_2O, H_2SO_4,} CH_3CHO$

(d) $HC\equiv CH \xrightarrow[Ni]{H_2,} CH_3-CH_3$

Ethane has a longer carbon–carbon bond than acetylene.

3. (a) $CH_3CH_2C\equiv CH$ 3 signals in ^1H-n.m.r. spectrum
                           4 signals in ^{13}C-n.m.r. spectrum
     $CH_3C\equiv CCH_3$ 1 signal in ^1H-n.m.r. spectrum
                           2 signals in ^{13}C-n.m.r. spectrum

(b) $CH_3CH_2C\equiv CH \xrightarrow[NH_3]{AgNO_3,}$ grey precipitate of $CH_3CH_2C\equiv C^-$
                                                     Ag^+

                   $\xrightarrow[NH_3]{CuCl,}$ reddish precipitate of $CH_3CH_2C\equiv C^-$
                                                     Cu^+

     $CH_3C\equiv CCH_3$ No precipitates formed with these reagents

4. (a) and (b) $HC\equiv CH \xrightarrow[HgSO_4]{H_2O, H_2SO_4,} CH_3CHO \xrightarrow{HCN} CH_3-\overset{\overset{\displaystyle H}{|}}{\underset{\underset{\displaystyle CN}{|}}{C}}-OH$

               $K_2Cr_2O_7\swarrow$ $\downarrow H^+, H_2O$

         CH_3COOH (a)

                                  $CH_3CHCOOH$
                                       $|$
                                  OH (b)

(c) $HC{\equiv}CH \xrightarrow{NaNH_2} HC{\equiv}C^-\overset{+}{Na} \xrightarrow{BrCH_2CH_2CH_3} HC{\equiv}CCH_2CH_2CH_3$

5. 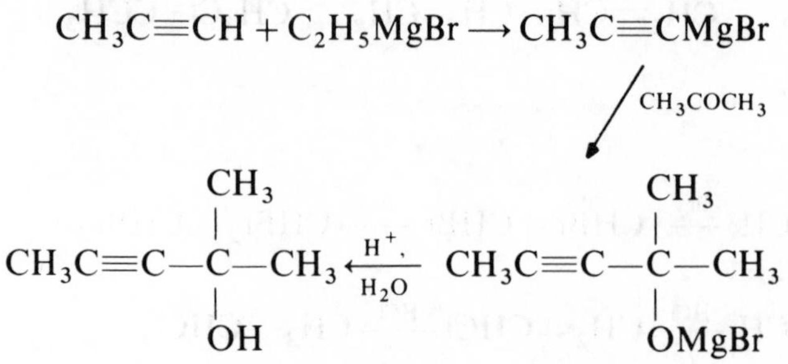 $CH_3C{\equiv}CH + C_2H_5MgBr \longrightarrow CH_3C{\equiv}CMgBr$

$$\overset{\diagdown CH_3COCH_3}{}$$

$$\underset{OH}{\overset{CH_3}{CH_3C{\equiv}C{-}\overset{|}{\underset{|}{C}}{-}CH_3}} \xleftarrow[H_2O]{H^+,} \underset{OMgBr}{\overset{CH_3}{CH_3C{\equiv}C{-}\overset{|}{\underset{|}{C}}{-}CH_3}}$$

Chapter 23

1. 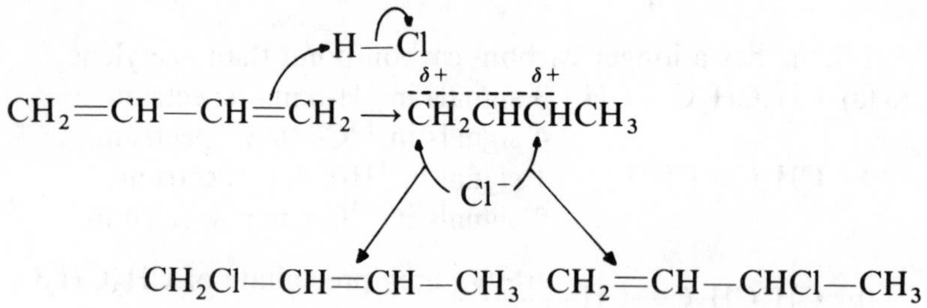 $CH_2{=}CH{-}CH{=}CH_2 \to \overset{\delta+}{CH_2}\overset{}{CH}\overset{\delta+}{CH}CH_3$ $H{-}Cl$

   Cl^-

   $CH_2Cl{-}CH{=}CH{-}CH_3 \quad CH_2{=}CH{-}CHCl{-}CH_3$

2. (a) $CH_3CH{=}CH{-}CH{=}CHCH_3 + 1Br_2 \longrightarrow$

$$\begin{bmatrix} CH_3CHBr{-}CHBr{-}CH{=}CHCH_3 \\ 4,5\text{-Dibromohex-2-ene} \\ + \\ CH_3CHBr{-}CH{=}CH{-}CHBr{-}CH_3 \\ 2,5\text{-Dibromohex-3-ene} \end{bmatrix}$$

$\diagdown Br_2$

$CH_3CHBr{-}CHBr{-}CHBr{-}CHBr{-}CH_3$
2,3,4,5-Tetrabromohexane

(b)

Hexa-2, 5-diene would **not** react with but-3-en-2-one because this diene is not a conjugated diene.

Chapter 24

1.

(a)   (b)   (c)   (d)   (e)

2.

o-Xylene   m-Xylene   p-Xylene

In ^{13}C-n.m.r. spectra:

4 signals   5 signals   3 signals

Chapter 25

1. (a) $C_6H_6 \xrightarrow[\text{conc. } H_2SO_4]{\text{conc. } HNO_3,} C_6H_5NO_2$

(b) $C_6H_6 \xrightarrow{\text{conc. } H_2SO_4} C_6H_5SO_3H$

(c) $C_6H_6 \xrightarrow[\text{AlCl}_3]{\text{CH}_3\text{COCl,}} C_6H_5COCH_3$

(d) $C_6H_6 \xrightarrow{D_2SO_4} C_6H_5D$

2. (a) $Br\!-\!Br + FeBr_3 \longrightarrow Br^+ + [FeBr_4]^-$

(b) $RCOCl + AlCl_3 \rightleftharpoons [RCO]^+ [AlCl_4]^-$

Chapter 26

1.

2. Most reactive: methoxybenzene
   Electron-donating substituent increases reactivity towards electrophiles.
   Least reactive: benzaldehyde
   Electron-withdrawing substituent decreases reactivity towards electrophiles.

(a)

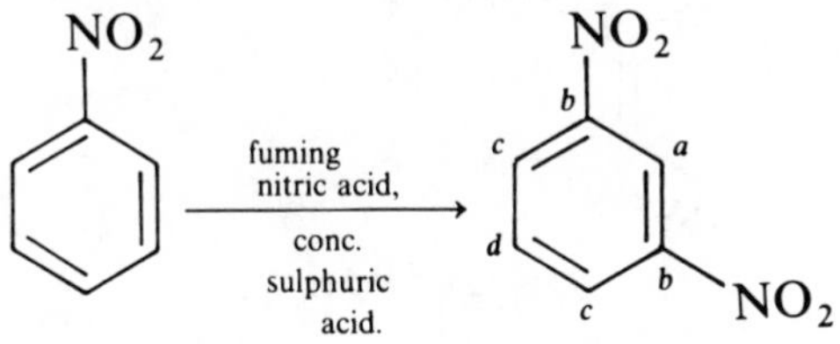

4 signals in ^{13}C-n.m.r. spectrum
(a–d)

(b)

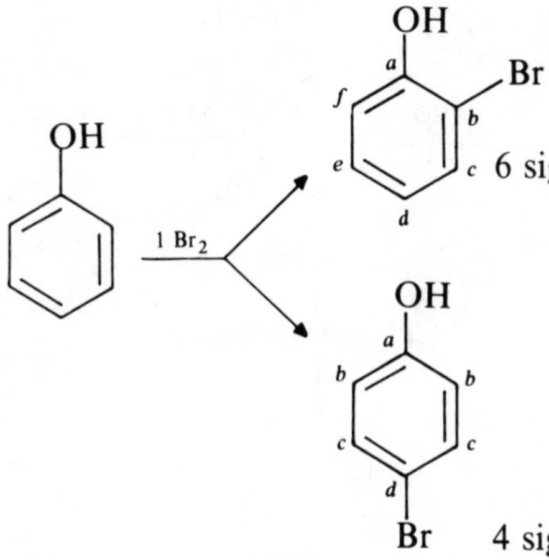

6 signals in ^{13}C-n.m.r. spectrum
(a–f)

4 signals in ^{13}C-n.m.r. spectrum
(a–d)

(c)

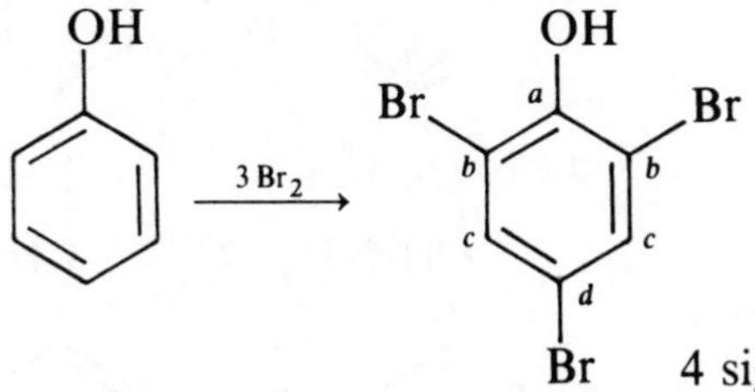

4 signals in ^{13}C-n.m.r. spectrum
(a–d)

Chapter 27

1.

2.

3.

4.

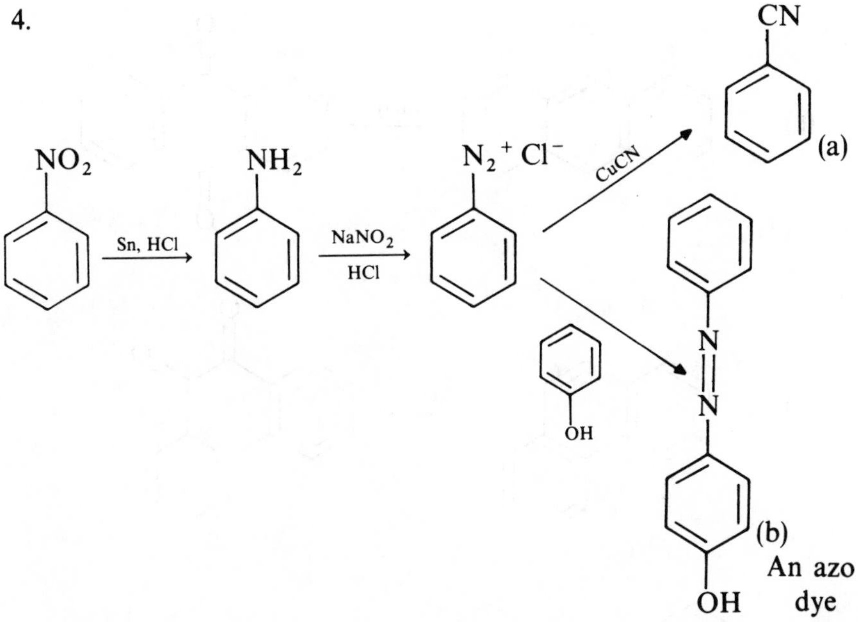

Chapter 28

1. (a)

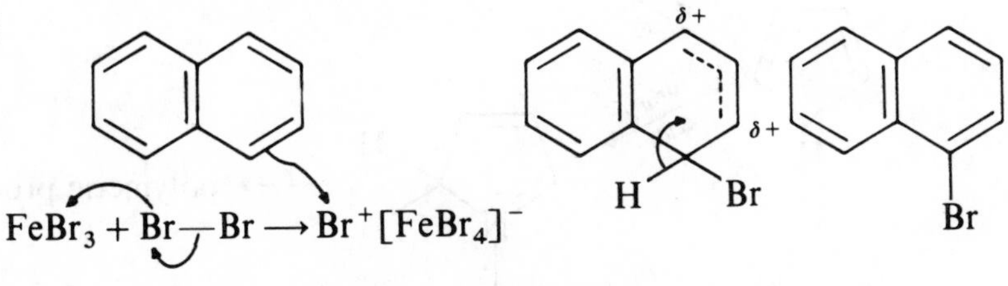

(b)

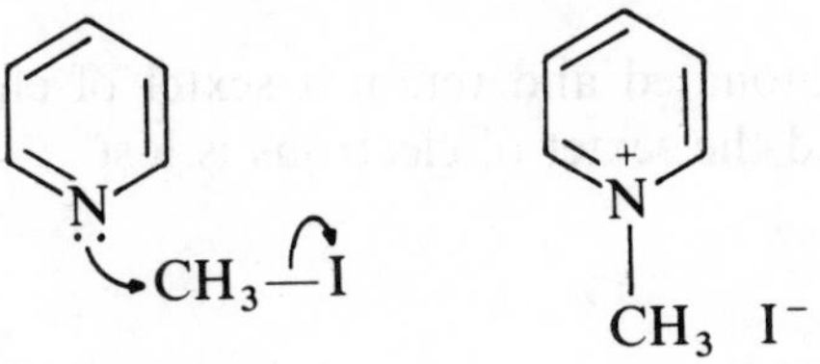

2. (a)

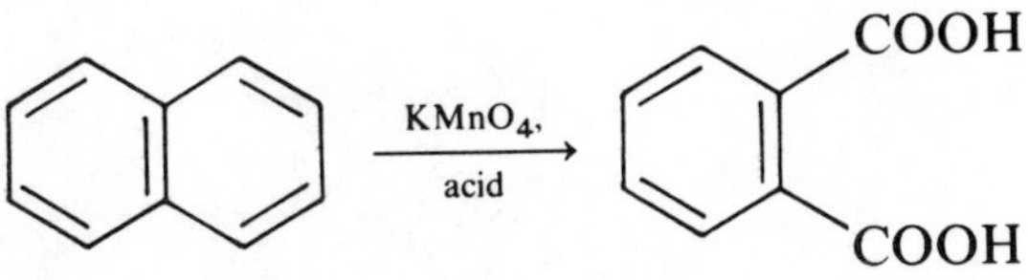

(b)

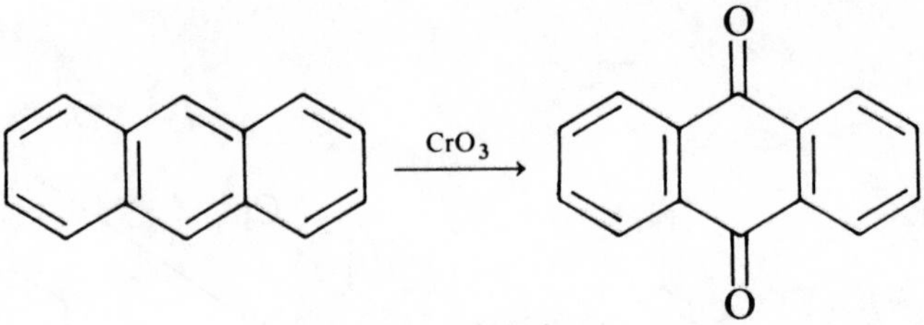

(c)

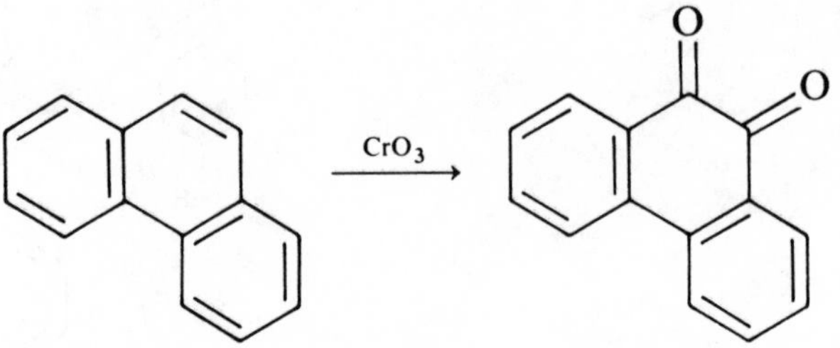

3. (a)

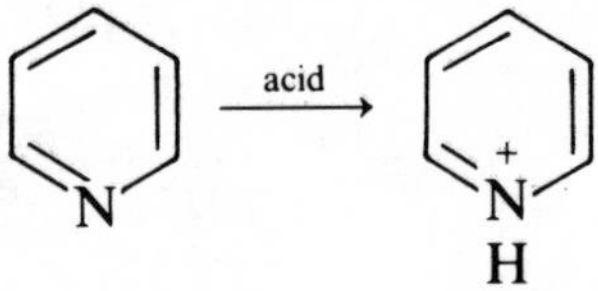

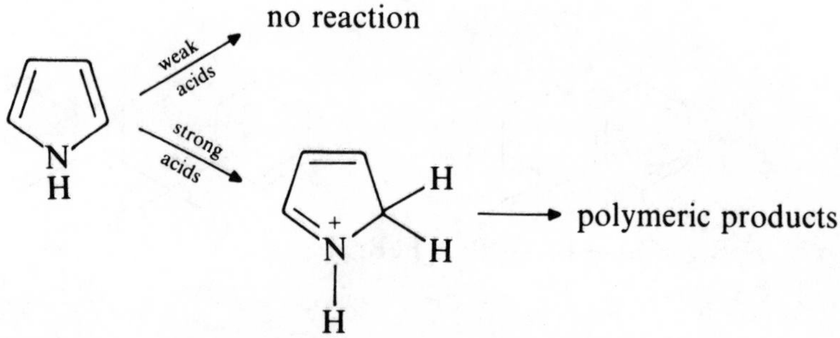

no reaction

polymeric products

Pyridine can be protonated and retain a sextet of electrons. When pyrrole is protonated the sextet of electrons is lost.

INDEX

Bold type indicates a more important entry